W0193133

Systems Thinking: From Heresy to Practice

Systems Thinking: From Heresy to Practice

Public and Private Sector Studies

Edited by

Keivan Zokaei
Consultant and Honorary Fellow at Warwick University

John Seddon
Consultant and Visiting Professor, Vanguard / Cardiff and Derby Universities

and

Brendan O'Donovan

Selection and editorial content © Keivan Zokaei, John Seddon and Brendan O'Donovan 2011
Individual chapters © the contributors 2011
Foreword © Prof. Michael C. Jackson 2011

All rights reserved. No reproduction, copy or transmission of this publication may be made without written permission.

No portion of this publication may be reproduced, copied or transmitted save with written permission or in accordance with the provisions of the Copyright, Designs and Patents Act 1988, or under the terms of any licence permitting limited copying issued by the Copyright Licensing Agency, Saffron House, 6-10 Kirby Street, London EC1N 8TS.

Any person who does any unauthorized act in relation to this publication may be liable to criminal prosecution and civil claims for damages.

The authors have asserted their rights to be identified as the authors of this work in accordance with the Copyright, Designs and Patents Act 1988.

First published 2011 by
PALGRAVE MACMILLAN

Palgrave Macmillan in the UK is an imprint of Macmillan Publishers Limited, registered in England, company number 785998, of Houndmills, Basingstoke, Hampshire RG21 6XS.

Palgrave Macmillan in the US is a division of St Martin's Press LLC, 175 Fifth Avenue, New York, NY 10010.

Palgrave Macmillan is the global academic imprint of the above companies and has companies and representatives throughout the world.

Palgrave® and Macmillan® are registered trademarks in the United States, the United Kingdom, Europe and other countries.

ISBN 978-0-230-28555-2 hardback

This book is printed on paper suitable for recycling and made from fully managed and sustained forest sources. Logging, pulping and manufacturing processes are expected to conform to the environmental regulations of the country of origin.

A catalogue record for this book is available from the British Library.

A catalog record for this book is available from the Library of Congress.

10 9 8 7 6 5 4 3 2 1
20 19 18 17 16 15 14 13 12 11

To Javad who is the greatest systems thinker I have known.
To Melanie who has been a great inspiration in completing this book.

KZ

Contents

Part II Systems Thinking in Private Sector

Illustrations

Tables

Figures

Foreword

The distinctive approach to service improvement discussed in this book is systemic and it drives improvement on the customer's terms. These two features are also what lead to the successful outcomes reported in many of the case studies.

John Seddon developed a version of systems thinking from a study of the Toyota Production System, which he regards as the best private sector example of systems thinking in practice, together with other concepts derived directly from the systems literature such as requisite variety. The approach was suitably modified to take account of the peculiarities of service systems. The systems ideas most clearly reflected in these case studies are:

- help managers to think differently (systemically)
- ensure that those doing the work lead the process of system diagnosis and redesign
- design the whole system to serve the customer's purpose
- connect all the elements of the system to contribute to serving customer purpose
- eliminate waste – those aspects of the process that do not contribute to customer purpose
- understand the type and frequency of demand on the system
- design against demand; increase the capacity of the system to respond by giving greater autonomy to those doing the work
- design support systems such as IT systems, only after the main operational system has been designed and ensure they support that system
- evaluate in terms of whether and how the whole system performs in meeting customer purpose
- provide for the continuous learning of those doing the work and their managers

Combined in the systems thinking 'Check-Plan-Do' methodology, which first exposes the reality of the current system's performance, these powerful ideas are capable of increasing effectiveness – doing the right thing by meeting the customer's purpose; increasing efficiency – often in ways that are counterintuitive; and improving employee morale. The case studies also demonstrate how reductionist approaches, driven by command and control thinking and embedded in regulatory and inspection regimes, targets and lean tools, inevitably make things worse. They encourage systems to do the wrong thing, driving up costs and damaging morale.

The second distinctive feature in the case studies presented in this book is the commitment to customer purpose. Systems thinkers bring the bad news to men and women of action that everything is interconnected; therefore it is very difficult to know whether a change made in one part of a system will make the whole system better. Systems thinkers narrowly escape being shot because they are willing to draw a boundary around their enquiries in order to get things done. They insist, however, that when a boundary is drawn it is explicit and that there is reflection upon the nature of the boundary and what elements and viewpoints it includes and excludes. The systems thinking approach – discussed in this book – draws its boundary by including those elements and perspectives that serve the customer's purpose, and is wholly explicit about this. In this respect, the editors are right to identify it with 'hard systems thinking', although it is a very sophisticated example. Systems thinkers of other persuasions will argue that there are circumstances when the boundary is appropriately drawn broader, to incorporate consideration of other processes that might best be redesigned alongside the process of choice or other stakeholder purposes. Be that as it may, as these case studies show, drawing the boundary and taking action based on a strong definition of customer purpose is not a bad place to start. It may be the only place to start when radical change is so urgently needed to protect and improve the performance of our public services in the face of severe financial cuts.

Professor Michael C. Jackson
October 2010

Contributors

Editors

Keivan Zokaei is a management consultant, researcher, and systems thinker. Currently he is an Honorary Fellow at Warwick University and Head of Operational Research at SA Partners. He is also a founding editorial board member of the *Lean Management Journal*. Previously he was Director of MSc in Lean Operations Management (Service) at the Lean Enterprise Research Centre (LERC), Cardiff University. Keivan has authored numerous articles on lean thinking as well as various reports for government departments and industrial bodies on lean and systems thinking. He has appeared on BBC Radio and acted as an advisor to the Auditor General for Wales. Keivan regularly delivers short course training to executives from different industries. His areas of interest are systems thinking, operations management, and lean and green.

John Seddon is an occupational psychologist, management thinker, and leading authority on change in organisations. He is a vociferous critic of centralised 'command and control' management of the public sector. John insists managers will discover they would be far better off if they knew how to manage for flow rather than aiming to create economies of scale. He is credited with translating the principles behind the Toyota Production System (TPS) for service organisations. John is a Visiting Professor at Cardiff and Derby Universities, and is managing director of Vanguard Consulting. He is author of *Systems Thinking in the Public Sector: the Failure of the Reform Regime and a Manifesto for a Better Way* (Triarchy Press, 2008).

Brendan O'Donovan is Head of Research at Vanguard Consulting, working alongside John Seddon. He has recently completed an MSc in Operations Management at Cardiff Business School. Having previously worked for a city council, Brendan has undertaken research projects for Cardiff University and recently co-authored a report for the Wales Audit Office on the subject of systems thinking in the Welsh public sector.

Contributors

Chris J. Backhouse, currently Director of Internationalisation Strategy at Loughborough University, received his BSc in Engineering Science from Durham in 1975, his MSc in Machines and Mechanisms from Liverpool Polytechnic in 1979, and his PhD in Spatial Differential Geometry from Liverpool Polytechnic jointly with Unilever Research Port Sunlight Laboratories in 1989. He was a Design Consultant in Liverpool Polytechnic from 1980 to 1982, and Scientist in Unilever Research from 1983 to 1990. He was also a Visiting Industrial Fellow at Cambridge University Engineering Department, Manufacturing Group from 1987 to 1989. He was a Design Manager at Morris Cranes and Senior Lecturer (joint appointment with Morris Cranes) at Loughborough University from 1990 to 1994. He was the Head of Dept of Manufacturing Engineering from 1995 to 1999, Director of Wolfson School of Mechanical and Manufacturing Engineering from 1999 to 2001 and Dean of Engineering from 2001 to 2007.

Barry Evans' early career involved a variety of roles in logistics/distribution with Watney Mann, Rank Hovis McDougall, and the Royal Mail. This was followed by roles in Tesco plc, including Lean Process Manager in Tesco Supply Chain Development. His role was to understand lean thinking and develop ways in which it could be applied in Tesco. He joined the Lean Enterprise Research Centre (part of Cardiff Business School) in 2005 as a research associate and has worked on projects for a range of clients related to lean thinking, value stream and value chain analysis, including DEFRA, the Wales Audit Office and the Food Chain Centre. He has co-authored a book on Value Chain Management: *Value Chain Management—Developing a More Valuable and Certain Future*, (with Tom McGuffog, Barry Evans, Peter Jordan, Jeremy Clarke, and Nick Wadsley; CILT 2009).

Rhian Hamer began her career as a graduate management trainee with the global steel maker, Corus. Fast-track progression enabled her to gain experience in all areas of the business from operations to marketing. Rhian trained as a Six Sigma Master Black Belt, leading quality and efficiency projects. After joining Her Majesty's Prison Services' Shared Services, Rhian enjoyed the challenges of implementing a systems thinking approach in a public sector transactional environment. In December 2009 Rhian moved to the Strategy and Change Directorate of the Ministry of Justice. Rhian has recently completed an MBA with Cardiff Business School specialising in Change Management, with particular emphasis upon sustainable change within the public sector. Her research in this area continues to inform and inspire her work in the field.

Julie Hilton is a Principal Tutor in Services Marketing, and a Senior Teaching Fellow, at the University of Derby. She has extensive teaching and management experience and has published several papers in the areas of Services Marketing and Retailing.

Mark Hopkinson has over 12 years of experience in improvement within complex supply chains. He has worked with companies ranging from aerospace electronics to precision foundries and machinists, and from call centres to civil contractors. He has an extensive knowledge of systems thinking and lean, and has applied these methodologies to improve performance both within companies and along their supply chains. Mark obtained an MSc in Lean Operations Management and a BEng Hons in Integrated Engineering.

Ayham A. M. Jaaron is currently a full-time Assistant Professor at the Industrial Engineering Department of An-Najah National University, Nablus, Palestine. He received his PhD in Manufacturing Engineering and Operations Management from the Wolfson School of Mechanical and Manufacturing Engineering, Loughborough University, UK in June 2010. He also received his BSc in Mechanical Engineering and MSc in Engineering Management from the University of Engineering and Technology (UET-Lahore) in 2002 and 2004 respectively. He was an Academic Visitor at the Design, Manufacture and Engineering Management Department (DMEM) of the University of Strathclyde, Glasgow, UK in 2006.

Michael C. Jackson is Professor of Management Systems and Dean of Hull University Business School. He is past President of the UK Systems Society, the International Federation for Systems Research and the International Society for the Systems Sciences. He has also served on the Council of the Operational Research Society. He is a Fellow of the British Computer Society, the Chartered Management Institute, the Cybernetic Society and the Operational Research Society. He is a Visiting Professor at the Indian Institute of Technology (New Delhi) and an Honorary Professor at the Universidad Ricardo Palma, Lima, Peru.

Sarah Lethbridge has worked in the Health and Services group at the Lean Enterprise Research Centre since 2005. She has led and participated in many lean projects within hospitals, public sector organisations, and local and national government. Her research involves understanding how to develop entrepreneurial cultures within organisations by exploring how to best transfer improvement knowledge to key internal change agents. She developed the successful Lean Skills for Managers programme within the Cardiff Lean University project which is training senior staff members on how to lead their staff through transformation projects. Recently, Sarah has been working with various departments in the Ministry of Justice to develop their lean learning programmes.

David Longbottom is Reader in Internal Marketing and the Director of the Systems Thinking and Organisational Change Research Group at the University of Derby. David has written and published over 40 academic papers and has presented at many international conferences in the areas of Total Quality Management, Business Excellence, Benchmarking, Best Practice, and Internal Marketing. His research in recent years has focused on the services sector, and in particular financial services. Before moving into research and academia David worked within the banking sector for over 20 years. David is a member of several academic and professional bodies; he is a Fellow of the Institute of Financial Services, an Associate of the Chartered Institute of Secretaries, and a member of the European Centre for TQM. He is a reviewer for several international journals.

Joe Marshall's teaching and research has focused on organisational change for a number of years at the University of Derby, and he was the leader of the part-time MBA programme at Derbyshire Business School. Joe has recently published research on the effects of organisational change on staff and patient care in the NHS and on organisational learning and the impact on performance in the manufacturing sector. His current research is on systems thinking and organisational change in the service sector.

Peter Middleton is a Senior Lecturer in Computer Science, Queen's University Belfast. His book *Lean Software Strategies* (Productivity Press, 2005) won the American 2007 Shingo Prize for Applied Research Excellence. His book of case studies, *Delivering Public Services that Work* (Triarchy Press, 2010), provides compelling evidence of the benefits from implementing systems thinking in the public sector. He received his PhD in computer science from Imperial College, London, and an MBA from the University of Ulster. Peter's research interests are applied systems thinking and lean software development.

Donna Samuel is currently completing her PhD thesis concerned with the diffusion of lean over time. Although her background is manufacturing, her more recent research interest is the application of lean in non-traditional environments such as the service and public sectors. She has worked for many years at the Lean Enterprise Research Centre in a variety of research and teaching roles and has published extensively.

Justin Watts is the Continuous Improvement manager at Burton's Foods. He is currently responsible for setting up lean principles, behaviours and results for the Burton's Foods Group. He previously worked implementing lean and Six Sigma programmes for a paper and packaging company, being trained in improvement methods as a supplier to Sony and LG Electronics. Engaged in rolling out the company's lean programme across 280 European manufacturing sites, Justin was seconded to Brussels for two years. Justin obtained an MSc in Lean Operations from Cardiff University's Lean Enterprise Research Centre, which was where he learnt more about the Theory of Constraints, Factory Physics, and Systems Thinking.

Ying Xia-Zheng (Vanessa) is a PhD student at the University of Surrey, and also an associate lecturer. Her PhD research is involved with investigating the role of mobile technologies on developing customer relationships. She has published several papers and presented at international accredited conferences, most recently at the Academy of Marketing.

1
Introduction

Keivan Zokaei, John Seddon and Brendan O'Donovan

The objective of this book is to enhance the understanding of service management concepts and their applications to service design. Arguably, little progress has been made since Ted Levitt's influential *Harvard Business Review* article 'Production-Line Approach to Service' was published in 1972. Since then, it has been common for services to be treated like production lines in the academic literature and more widely in management practice. However, the belief that achieving economies of scale will reduce unit costs has been a common feature of management decision-making.

One of the authors (John Seddon claims to have translated the principles behind the famous Toyota Production System and made them applicable for service organisations (Seddon, 2003). The Toyota Production System is just that – a system designed to produce cars at the rate of demand. Its originator, Taiichi Ohno, realised a number of counter-intuitive truths, ideas that challenged conventional management wisdom. When service organisations are studied as systems, they reveal similar counter-intuitive phenomena.

This book aims to:

- explore the application of systems thinking to a variety of public and private sector service organisations;
- draw distinctions between conventional approaches to change in both service and manufacturing organisations and those using systems thinking methods;
- illustrate the counter-intuitive truths revealed by studying service organisations as systems; and
- illustrate with case studies how organisations have been redesigned using systems principles, resulting in a strong impact on performance and morale.

1

We have compiled chapters from practitioners and management theorists working in operations management on the topic of systems thinking in service organisations.

Background of systems thinking

Until the middle of the twentieth century, the dominant method of scientific inquiry into an organised system was to reduce it into separate elements, and to study each element individually. Underlying this reductionist approach was the notion that the whole is no more than the total sum of its parts. However, during the 1930s and 1940s, both scientists (von Bertalanffy, 1940, 1950; Wiener, 1948) and philosophers (Fries, 1936) gradually realised that a complete understanding of a system required holistic study of the individual constituents of the system as well as their inter-linkages and the relationships with the wider system. Underlying this systemic approach is the idea that additional characteristics emanate from the whole which are not attributable to any particular part of the system; in other words, the system is more than just the total sum of its parts. Systems thinking then is 'the scientific exploration of "wholes" and "wholeness" which, not so long ago, were considered metaphysical notions transcending the boundaries of science' (von Bertalanffy, 1972, p. xviii). Flood (1999) states that 'we can only meaningfully understand ourselves by contemplating the whole of which we are an integral part.'

Although initially introduced in science, systems theorists soon extended their organismic metaphor to include social and man-made entities. As such, systems thinking diffused into disciplines such as general management (Deming, 1982; Ackoff, 1971), logistics (Forrester, 1958), cybernetics (Wiener, 1948), and most recently in the service industry (Seddon, 2003). Systems thinkers in service industries have also drawn upon the ideas developed in manufacturing in Japan after World War II, especially those of influential thinkers such as Deming (Deming, 1982) and Ohno (1988) (Seddon 2003, ODPM 2005).

Deming was a distinguished management advisor to the Japanese during the American post-war reconstruction of their economy. He argued that Western organisations and thus Western economies were in crisis because of their beliefs in flawed management assumptions. 'Most people imagine that the present style of management has always existed, and is a fixture. Actually, it is a modern invention – a prison created by the way in which people interact' (Deming, 1994).

Deming's point was simple: mankind invented management, therefore mankind could reinvent it. His work included a scathing and detailed critique of Western management assumptions. The main targets for criticism were the use of arbitrary measures to govern the way work is managed, the management of separated functions independently within an organisation, and the separation of decision making from the worker. The better alternative, he argued, was to understand and manage organisations as systems.

Taiichi Ohno and the Toyota Production System (TPS)

The tale of the superior performance of Toyota over its mass-producing competitors was first brought to widespread Western attention by *The Machine that Changed the World* (Womack, Jones and Roos 2007). First published in 1990, Womack, Jones, and Roos' book used the term 'lean' to describe what had occurred at Toyota. Through experimenting first with simple die-change techniques (ways of stamping metal sheets), Toyota discovered that they were able to perfect the whole process until it was reduced from taking one day down to only three minutes. In doing this, Ohno made the first of a series of counter-intuitive discoveries: it cost less per part to make small batches of stampings than to produce them in large batches. Seddon and Caulkin (2007) saw these discoveries as Ohno realising over time that economy of flow was superior to economy of scale, and that, in order to see flow, he needed to understand his organisation as a system: 'Ohno's creation, the Toyota production system (TPS), is the most strikingly successful example of systems thinking applied to business organization.' Ohno, the architect of the TPS, was thus claimed by the authors as a systems thinker.

The belief that achieving economies of scale will reduce the unit cost is a common feature of management decision-making which ignores the innovations of Ohno and the TPS. As technological advances have produced more sophisticated IT and telephony, it has become easier for conventionally-run firms to standardise and off-shore their services in the pursuit of lower costs. The development of the 'lean' movement in the years following Womack and Jones' initial popularisation of the term has only helped to emphasise the same underlying management assumptions: by managing cost and workers' activity, organisational performance is expected to improve. Through misinterpretation of the core paradigm of management, 'lean' has become subsumed into the 'business as usual' of conventional service management. As a result, 'lean' has become synonymous with 'process efficiency' and the opportunity for significant performance improvement – as exemplified by Toyota – has been missed.

Taken as a whole, an underlying theme emerges from the studies included in this book that the focus for the design and management of services needs to be on *delivering effectiveness*, which can be achieved by providing exactly what the end customer wants; no more and no less. If a system is designed against this ideal, it can continually perfect the way it works to deliver value to the end customer. In transactional systems, this is achieved by having an excellent understanding of the demands being placed on a service by customers and then by continually improving the way the system delivers against these demands. In the public sector, through studying demand and the flow of work through a system, it is possible to discover that there may be services for which there are no demands, and yet there may be other demands which an organisation does not currently cater to, but may indirectly affect another public service if it is not dealt with (e.g. demands of older people on Adult Social Care services which are not dealt with and instead become acute health demands on the NHS). This also offers an indirect route to efficiency: do what matters to a customer, get it right the first time, and operations will become optimally efficient.

Seddon (2003) proposes a systems thinking approach which is more akin to hard system methods where the system is assumed to be identified with a single unifying purpose. Some research suggests that this approach has found success in the UK public and third sectors (OPDM, 2005; Advice UK, 2008). Seddon distinguishes between two approaches to management and the design of work: conventional approaches (which he terms 'command and control') where fragments of an organisation are optimised with little reference to the wider organisation; and a systems approach which focuses on the interrelationship between the various parts of the organisation.

Command and control is defined as 'regulation by management, with its battery of computer and other informational aids...where decision-making is distant from the work and based on abstracted measures, budgets and plans' (Seddon and Caulkin 2007). Systems thinking emphasises not just 'wholeness,' but also the 'thinking of the system' (i.e. that of the managers and workers within a system) which needs to change in order for the system to be able to improve. Table 1.1 shows some of the key differences between the two approaches.

Many chapters in this book report on methods developed for applying systems thinking in service organisations. The application of systems thinking following Seddon's Vanguard Method uses a 'Check-Plan-Do'

Table 1.1 Command-and-Control thinking vs. systems thinking

Command-and-Control thinking		Systems thinking
Top-down	Perspective	Outside-in
Functional specialisation	Design	Demand, value, and flow
Separated from work	Decision-making	Integrated with work
Budget, targets, standards, activity and productivity	Measurement	Designed against purpose, demonstrate variation
Extrinsic	Motivation	Intrinsic
Manage budgets and people	Management ethic	Act on the system
Contractual	Attitude to customers	What matters ... ?
Contractual	Attitude to suppliers	Partnering and cooperation
Change by project/ initiative	Approach to change	Adaptive, integral

cycle (adapted from the Plan-Do-Check-Act cycle recommended by Deming). The Check phase of the improvement provides a framework for understanding and knowledge about the system.

Moreover, it is designed so that the thinking of the participants is changed during the analysis (Seddon 2005). 'Check' opens the eyes of the organisation to the 'command-and-control' principles and philosophy which are underpinning the design of the current system. It is these command-and-control principles which cause the suboptimal performance of this system. Thinking needs to be changed before acting on the system and achieving improved performance.

Another key feature in the following chapters is the issue of targets and measures in the service sector, and particularly in the public sector. Many of the cases show how imposing arbitrary measures in the shape of targets and standards create a de facto purpose (i.e. to meet the targets). Elsewhere, the literature has covered the numerous unintended consequences of centrally set targets, and their limitations in the public sector (Bevan and Hood 2006). The requirement to 'meet the targets at all costs' leads to the instances of so-called 'gaming.' becoming myriad. Also, by specifying how services should be run against standardised models, the ability of local service providers to innovate and continually improve is removed (i.e. 'dumbed-down' standardisation).

This brief introduction considers the role of broader systems thinking in the management literature, the development and innovations which emerged in the Toyota Production System, and the translation of these principles for their application in service systems. Many of the chapters delve deeper into the methods involved.

An overview

This book consists of two parts. Part I, which follows this introduction, consists of seven chapters and each presents a different case or different aspect of applying systems thinking in the public sector from Local Authorities to Housing Associations to Shared Services. Part II includes four chapters written by experts who look at the application of systems thinking in the service industry in the private sector ranging from utilities to banks to software development. The last chapter in Part II is a case study from the manufacturing sector. On the surface, this may be considered an odd addition for a book that focusses on service management, but the last chapter is valuable in that it refers to some of the fundamental building blocks of operations management that are highly relevant to both manufacturing and service industries.

Part I – Public sector case studies

In Chapter 2, Zokaei discusses how systems thinking can provide a framework for change by illustrating some of its key features deployed in the redesign of a social service system. Zokaei presents a case study of a Disabled Facilities Grant service in a Local Authority emphasising the following key aspects of the systems thinking intervention:

- In systems thinking the emphasis is on 'effectiveness thinking' as opposed to 'efficiency thinking.' Effectiveness is described as doing the right thing and efficiency as doing things right. Effectiveness thinking is concerned with quality and service, whereas efficiency thinking is concerned with cost and activity measurement. As one of the leaders of the service studied in Chapter 2 put it, 'if we focus on delivering service to citizens in the process of doing so we also become more efficient.' This is a counter-intuitive moment; yet all managers who have led various systems thinking interventions presented in this book at some point allude to this principle. This principle is especially important at a time that public services are facing great pressure to deliver efficiencies. Senior management in the public sector should realise that by focussing the improvement initiatives on cost, they are likely to increase costs. On the other hand, by focussing on delivering the purpose from citizens' view points, they are likely to reap great benefits as described in all seven public sector case studies in this book.

- In systems thinking, improvement always begins by understanding the work (i.e. seeing the real thing in the workplace). In order to be truly effective managers, champions should be intimately involved with the system they are trying to improve. Improvements are therefore work based and bottom-up as opposed to the all too common approach of 'mandating best practices' in public services, which rarely works.
- Action learning is at the heart of systems thinking. The author explains that action learning not only ensures that solutions are work based and context rich as discussed above, but also helps people to surface their underlying assumptions and change them after experiencing the (often negative) effects of their own unconscious values. This is referred to as normative education and is deeply rooted in the works of systems thinkers such as Argyris (1999) who call for unlearning through hands-on experimentation before deciding upon new solutions.

The case study in Chapter 2 illustrates how applying these underlying principles of systems thinking leads to significant improvements including:

- Reducing the average end-to-end time of the service from 675 days to 64 days. That is, more than a 90 percent improvement.
- Reducing the (activity) cost of delivering the service (i.e. staffing costs for administration) from £499 to £319. That is, a 36 percent improvement.
- Reducing the average cost of the physical works from £7000 to £6300. That is, a 10 percent improvement.
- Reducing failure demand from 71 percent to 40 percent.
- But most important, the case study shows that by providing service in time citizens could increase the length of stay at their own residence and delay transfer to care by 4 years.

In Chapter 3 O'Donovan shows how the perception of large demand for adult social care leads councils to screen out demand (i.e. applicants for the service) through the strict application of eligibility criteria. The author explains that underlying the 'screening out of applicants' is the assumption that if all demand is dealt with first time the system would run out of capacity quickly. O'Donovan presents a detailed case study of an adult social care service in an English Local Authority where screening is part of the service before the systems thinking intervention. He explains how the intervention helped to refocus the service on doing what matters to the end user 'right-first-time' and the removal of screening out procedures. The case study reports an 87 percent cost reduction, i.e. average cost of administration went down from £998 to £134 per case. Moreover, 74 percent of the demand under the old design was failure demand whereas less that 10 percent of the demand under the new design (post intervention) was failure demand. Interestingly the case study reports a 'perceptible drop in the overall demand' by more than 30 percent. Often, a concern associated with removing the screen out systems such as the one reported in Chapter 3, is that it would lead to a spike in demand. However, O'Donovan makes a case for understanding the demand better and trying to solve citizens' problems 'first-time-right' rather than screening out or otherwise delaying by the service on the 'assumption' of lack of capacity. The case study shows that the administrative cost of delivering the service was £998 where the actual service cost only £105. This was partly due to the extremely bureaucratic way in which the service was delivered. O'Donovan draws attention to Taguchi's concept of 'nominal value' and how it sits at the heart of Seddon's systems thinking intervention method. O'Donovan concludes that 'by focussing on the customer's nominal value and effectiveness (better service and an improved service user experience) systems thinking was able to deliver substantial efficiency improvements as a second order result. This is the same continuous improvement experience as witnessed in the Toyota Production System. By eradicating the requirement for "feeding the performance machine at all costs" from this system, there was less of what Taguchi would call the "loss to society" from not having delivered to the customer's nominal value.'

In Chapter 4, Hamer and Lethbridge address the issue of shared service centres in the public sector. There has been a lot of support for shared service centres driven by the notion that economy of scale lowers costs. System thinkers have long argued against the fallacies of the economy of scale. Taiichi Ohno, the father of Toyota Production System, said 'an increase in production volume shouldn't necessarily mean a decline in unit costs any more than a decline in volume should mean an increase in unit costs. Those sorts of things happen as the result of arranging things poorly' (Ohno quoted in Shook, 2010).

Hamer and Lethbridge provide some evidence illustrating that the shared service improvement programme did not necessarily deliver the promised benefits. The authors then discuss some of the reasons why many change initiatives don't deliver in public sector organisations. They demonstrate how focussing solely on meeting the Service Level Agreements which were not driven from customer needs lead to customer dissatisfaction, failure to deliver the real purpose, and employee disengagement. The authors conclude that 'dislocation from the customer and measurement systems focussed on optimising parts of the end-to-end process comprised a system condition which obstructed workers from being able to deliver the best services.' This is a work in progress and a new phase of the programme is being developed. The new programme will consider that a problem solving ethos and continuous improvement culture must be supported at every level of the organisation.

Chapter 5 presents a case study from Portsmouth Housing Association revealing how they worked with suppliers to deliver truly remarkable improvements. The case study demonstrates how Portsmouth Housing Association reduced end-to-end time of the repairs and restoration service by 71 percent (from on average 24 days to less than 7 days) and reduced failure demand by 75 percent (from 60 percent to 14 percent). Moreover the case study reports how an individual contractor reduced their cost per job by 56 percent, from £258 to £114 while increasing capacity by 265 percent (from 85 jobs per day on average to 225 jobs per day), allowing them to accept additional work from other contractors who were not willing to apply systems thinking principles (or improve as quickly) with the same number of staff. This case study of collaboration demonstrates the potential to rewrite the guidance on strategic partnerships, and to serve as the benchmark for economic performance in the public sector. In this chapter, O'Donovan and Zokaei, also explain a tale of organisational change and how the existing policies and procedures adopted by the Audit Commission failed to recognise the improvement delivered at Portsmouth.

The overall domain of research in Chapter 6 is organisational change; Marshall looks into the growing interest in systems thinking in the housing sector and explains how systems thinking brings about organisation change, improvement, and performance. He provides an overview of the underpinning philosophies of systems thinking contrasting it to the command and control philosophy which underlies conventional management. Marshall asserts that 'conventional management employs a command and control paradigm; is top-down and hierarchical, separated from the work, target and budget driven with an ethos of central control and reaction instead of learning and adaption.' The chapter presents and examines quantitative data related to the impact of a systems thinking intervention on organisation performance in a Housing Association, and presents and examines qualitative data on the perceptions, thinking, and behaviour of individuals and groups before, during, and after the intervention.

Marshall is particularly interested in understanding whether the changes in thinking, behaviour, service quality, and organisational performance are sustainable. He offers an evaluation framework which links the quantitative performance data to qualitative evidence about the perceptions of those (directly or indirectly) involved in the intervention. Marshall identifies some of the key influences and conditions that explain momentum, spread, and sustainability of the systems thinking intervention in the housing association in question. The chapter is a work in progress but ends by explaining that the results to date present a partial yet favourable picture of the sustainability and spread of systems thinking from a variety of different stakeholders' point of view, particularly the performance of service delivery in terms of customer purpose.

In Chapter 7, Evans and Samuel examine how effective the systems thinking approach has been in the housing sector in the UK. They provide an interesting review of the application of systems thinking in the housing sector which shows limited adoption but considerable success. They then provide two case studies of their own from Coastal Housing Association and Charter Housing Association. The case studies illustrate impressive improvements, for example in the Allocations and Lettings department at Charter Housing Association the rent arrears have reduced from 31 days to 14 days (a 55 percent improvement). In the same department, the complaints were reduced from 104 per year to just 2 per year. We recommend this chapter to anyone who wants overview of the growing adoption of systems methods in this sector.

In Chapter 8, Jaaron and Backhouse make a case for why 'focusing on systems effectiveness leads to better efficiency,' while the opposite is not necessarily true. Jaaron and Backhouse review the literature on the application of manufacturing paradigms in public services arguing that studies of call centre management practices show that there is a tendency to focus on efficiency rather than effectiveness and measuring quantity rather than quality. The authors examine the issue from an organisational commitment point of view and identify the key aspects of organisational commitment zooming in on Affective Commitment as the antecedent of organisational success.

Chapter 8 features a case study examining whether the alternative management models applied to manufacturing add value to the public service department (in terms of enhanced customer service and reduced overall operational cost) and help improve the *affective commitment* of their employees and thus improve their value to their company. The case study examines a systems thinking redesign in a call centre at Stockport Council carrying out 'before and after' analysis pertaining to employees' affective commitment. The researchers report 'a dramatic change in the philosophy of work as compared to the traditional office values found in majority of service departments in the government sector.' In addition, the case study illustrates:

- 85 percent of the demand is met and dealt with right-first-time after the systems thinking redesign against only 17 percent of the demand being met right-first-time beforehand.
- It took less than a day to fix a call post redesign versus 11 days for the same type of call before the intervention.

These are significant improvements, and the authors conclude that deployment of systems thinking in designing public services results in significant benefits for employees (especially *affective commitment* of front-line employees), managers, customers, and eventually the organisation.

Part II – Private sector case studies

In Chapter 9, Longbottom, Hilton, and Xia-Zheng investigate a series of business improvement programmes within a major UK bank. Beginning with a review of quality and marketing-based improvement programmes, the authors go on to show that most of the changes in this bank were made in line with established change models. Surveys reveal that many of these programmes were considered to be unsuccessful, so this chapter delves deeper into the reasons for this through interviews with participants. The authors then make some recommendations for alternative strategies, including focussing the strategic alignment on customer requirements, ensuring the integration of efforts across the organisational structure, and finally gaining a better understanding of value from a customer's perspective. As ever, the cost savings are an outcome of the process of focussing on customer needs (or whatever is the real purpose the system); not the focus of the intervention.

Chapter 10 begins with a high level overview of systems thinking theories and categorisation of various systems thinking intervention methodologies. In this context, Hopkinson gives an explanation of Seddon's Check-Plan-Do approach against 'counterpart' approaches such as Checkland's Soft System Methodology and Ackoff's Interactive planning. He then discusses different service system archetypes and considers the Check-Plan-Do approach to be an appropriate intervention method for the 'demand-driven' service studied throughout the chapter. He also provides a detailed case study of an intervention in a 'break-fix system' in an electricity network repair and restoration service. While the case study tells the story of the application of the Check-Plan-Do model, Hopkinson provides hypothetical comparisons with Checkland's and Ackoff's intervention approaches concluding that there are key similarities.

Chapter 10 identifies the potential strengths in each approach and emphasises the great deal of improvement attained through the Check-Plan-Do approach. We agree with the author that the benefits of such theoretical comparisons are somewhat limited and that the discussions remain 'academic.'

The chapter also provides an interesting case study of improving service and releasing cash in the utilities sector. The distribution network service studied in this chapter reduced cost by 24 percent, i.e. £14.5 million cost savings in 2008 in a service which was costing £60 million per annum. At the same time, the systems thinking intervention improved the average end-to-end response to customer demand by approximately 30 percent while reducing the upper control limit of the elapsed time of responding to demand by approximately 45 percent. Hopkinson provides a detailed study of what went on during the intervention describing moments of truth that helped the team redesigning the service against customer demand and requirements leading to the above mentioned benefits.

In his account of applying systems thinking in private services, Hopkinson provides a detailed description of how old measures drove wrong behaviour. He explains how the call centre staff were measured on items such as 'politeness in handling customer calls' and 'willingness to help' rather than 'how well the demand is dealt with right-first-time' or 'quality of data recorded for use by the team tasked with repairing the fault.' Clearly, the old measures served as negative system conditions and needed to be replaced with customer facing measures that helped to attain the system's purpose as explained by the author. We recommend this chapter to anyone interested in understanding the impacts of system conditions and the role of good measures in attainment of the purpose.

Chapter 11 investigates a systems approach to adopting new IT packages into an organisation. The literature review critiques current 'best practice' approaches such as Agile or CMMI as not allowing for a full understanding of an organisation's IT needs. It is argued that the systems approach, which begins with understanding a service's needs as a system before embarking on an IT change programme, allows for a more successful implementation of change. The authors conclude that those currently advocating Agile or plan-driven approaches need to widen their scope to a systems rather than just a software perspective.

Chapter 12 is unusual as it analyses a systems approach applied in a manufacturing environment (in paper-based packaging factories). The author uses conventional 'lean manufacturing tools' and techniques in a 'transformation' site, but compares the results achieved with those of a 'control' site where no tools were applied. Despite the success of the lean tools by their own criteria, the 'control' site had less variability and thus better flow. The author uses Factory Physics and the Theory of Constraints to analyse the results. Both concepts contribute significantly to the body of knowledge in service management and systems thinking.

References

Ackoff, R. (1971) 'Towards a System of Systems Concepts'. *Management Science*, 17(11), 661–671.

Advice UK (2008) *It's the System, Stupid! Radically Rethinking Advice*. Advice UK: London (Downloadable from www.adviceuk.org.uk, accessed 7/4/2009).

Advice UK (2009) *Interim Report: Radically Rethinking Advice Services in Nottingham*. Advice UK: London.

Argyris, C. (1999) *On Organisational Learning*. Blackwell Publishing: London.

Bevan, G. and Hood, C. (2006) 'Have Targets Improved Performance in the English NHS?' *British Medical Journal*, 332, 18 Feb.

Checkland, P. (1997) 'Systems' in International Encyclopaedia of Business and Management Thomson Business Press, cited in Chapman 2002.

Deming, W. E. (1982) *Out of the Crisis*. MIT Press: Massachusetts.

Deming, W. E. (1994) *The New Economics: For Industry, Government, Education*. MIT Press: Massachusetts.

Flood, R. (1999) *Rethinking the Fifth Discipline, Learning within the unknown*. Routledge.

Forrester, J. W. (1958) 'Industrial Dynamics: A Major Breakthrough for Decision Makers' *Harvard Business Review*, 36(4) (July–August), 37–66.

Fries, H. S. (1936) 'On an Empirical Criterion of Meaning'. *Philosophy of Science*, 3(2), 143–151.

Office of the Deputy Prime Minister (ODPM) (2005) *A Systematic Approach to Service Improvement Evaluating Systems Thinking in Housing*. ODPM publications: London.

Ohno, T. (1988) *Toyota Production System*. Productivity Press: Portland, Oregon. Translated from Japanese original, first published 1978.

Seddon, J. and Caulkin, S. (2007) 'Systems Thinking, Lean Production and Action Learning'.*Action Learning Research and Practice*, 4(1), April 2007, special issue: *Lean Thinking and Action Learning*.

Shook, J. (2010) *Forward to Fundamentals: Webinar with John Shook Written Transcript*, [online] http://www.lean.org/common/display/?o=970 (accessed 01/07/2010)

von Bertalanffy, L. (1940) 'Der Organismus als physikalisches System betrachtet (the organism considered as physical system)'. *Die Naturwissenschaften*, 28, 521–531.

von Bertalanffy, L. (1950) 'An outline of General Systems Theory', *British Journal for the* Philosophy of Science, Vol. 1, 139–164.

von Bertalanffy, L. (1972) 'Forward to Laszlo, E., *Introduction to Systems Philosophy: Toward a New Paradigm of Contemporary Thought*. Gordon and Breech Science Publications: London.

Wiener, N. (1948) *Cybernetics; or, Control and Communication in the Animal and the Machine*. Wiley: New York.

Womack, J. P., Jones, D. T. and Roos, D. (1990) *The Machine that Changed the World*. Macmillian: New York.

Womack, J. P., Jones, D. T. and Roos, D. (2007) *The Machine that Changed the World*. Macmillian: New York. First published 1990.

Part I
Systems Thinking in Public Sector

2
How Systems Thinking Provides a Framework for Change: A Case Study of Disabled Facilities Grant Service in Neath Port Talbot County Borough Council

Keivan Zokaei

This chapter discusses how 'systems thinking' provides a framework for change by illustrating some key features of the approach deployed in the redesign of a public service system. The Check-Plan-Do methodology (Seddon, 2005) provides a structured framework for understanding the system and a practical framework for redesign. In contrast with many other change methodologies, in the systems thinking approach, there are no toolkits to be applied to managers' problems, or training courses for managers to attend. Instead, participants are required to continue to follow the method and to ensure that they are engaged in the study of their service in a systematic way.

A key feature of the systems thinking approach is its emphasis on 'effectiveness thinking' as opposed to 'efficiency thinking.' All case studies presented in this book explain how services are refocused into concentrating on delivering against the real purpose of the system. By doing so, services realise benefits that often exceed managers expectations. In all case studies presented here the systems thinking intervention involves conscientious study of the demand in a 'longer than usual' process which then proves to be valuable in the redesign. Becoming intimately familiar with the customers and demand is at the heart of the systems thinking approach. Systems thinking presents a radical shift away from old ways of management thinking; as one of the leaders of the service studied in here puts it: 'if we focus on delivering service to citizens in the process of doing so we also become more efficient'.

Another feature of the approach is that workers themselves are responsible for the redesign of the system in which they work, which is a powerful way of engaging the workers. This approach to change, borrows from the works of Argyris (1999) and Argyris and Schön's (1974) theory of double-loop learning. Action learning is therefore at the heart of systems thinking. Improvements are work based and bottom-up as opposed to the all too common approach of 'mandating best practices' which hardly ever works.

But action learning is more than just a good method for finding appropriate, context rich solutions. Action learning (or 'normative' educative) helps people to surface their underlying assumptions and change them after experiencing the (often negative) effects of their own unconscious values. Argyris (1999, p. 90) suggests that 'it is possible to help individuals learn new theories-in-use and to create new learning systems. The intervention requires the creation of a dialectical learning process where the participants can continually compare their theories-in-use, and the learning system in which they are embedded, with alternative models. This requires that interventionists make available alternative models with significantly different governing values and behavioural strategies'. This point is explained further in the following.

The following case study provides an overview of the improvement process carried out in the Disabled Facilities Grant (DFG) service in Neath Port Talbot County Borough Council (henceforth NPT) and the progress made by the end of 2009. At the time of this study, the improvement approaches had reached a point of maturity whereupon staff continuously pursued better ways of working and therefore, by the time of publication, the measures of improvement could be over and above what is reported below.

Background

In November 2008 the NPT County Borough Council embarked on the review of the DFG service. The DFG 'systems thinking' improvement was part of expanding in-house learning and was fully supported by an external consultant. This was against a backdrop of increasing demand for disabled facilities grants and the perception that there was a need to expand capacity. One of the key aims of the review, inter alia, was to design a new system capable of responding to the high level of demand which reflected the national data on long-term disability, produced by the Office of National Statistics, indicating that approximately 34 percent of the population of Neath Port Talbot have a long-term disability.

In support of the demand for DFGs, between 2004 and 2009, NPT has invested £15 million in disabled facilities adaptations for private householders. A further £5.5 million has been spent on similar works for council tenants. Notwithstanding the high level of investment over this period, demand for adaptations has continued to grow. At the time of this study in late 2009, and despite significant improvement in the service, there are approximately 450 cases on the waiting list.

Moreover, the service was judged to be performing poorly against the national indicator for DFGs, which placed the council 21 out of 22 authorities on the average end-to-end time for delivery of a DFG. However, this conflicted with the annual customer satisfaction questionnaire for the same period which indicated that over 99 percent of customers were satisfied with the DFG they received.

Demand presents itself in DFG service via referrals and the Occupational Therapy (OT) service provides the first point of contact for DFG. Each application must be supported by a 'disability needs assessment', which is carried out by the OT at the start of the process. At the outset the aim of the review was:

- to maximise the benefit of the Council's investment in DFGs.
- to ensure the efficiency of the DFG service.

However, in a gradual process of learning and gaining a thorough understanding of the ability, or inability, of the current system to achieve customer needs, the objectives of the review were redefined as to design a new service that:

- is capable of responding to demand for DFGs.
- will have little waste, if any.
- is based on a continuous process of measurement, review, and improvement.
- trains staff to meet the demands on the service.
- focuses delivery on maximising independence at home.
- effectively uses resources to ensure that the purpose of the service is achieved.
- takes full advantage of partnership working with health, social care, social landlords, independent providers, and the third sector.

One important point is that the systems thinking review team realised that their role was to create capacity through efficiency, to support the high levels of demand evident from waiting list queues as opposed to simply delivering financial savings.

The improvement process

In November 2008 a team was put together to take DFGs through a systems thinking review. The multi-skilled team consisted of five officers each with considerable experience in the way DFG service operates. The team was led by the service manager and was fully supported by senior leaders who also got involved in experiencing the nature of the work on the frontline. This involved senior leaders participating in frontline activities such as listening to demand, speaking to staff members about their work, asking 'what stopped them doing a good job', and asking customers 'what mattered' to them.

The improvement started with the external consultant training the senior leaders and team members in the principles of lean and systems thinking. The team followed the systems thinking method which consists of three steps as outlined in the following diagram. Figure 2.1 and Table 2.1 provide explanations and the timeline of each step of improvements.

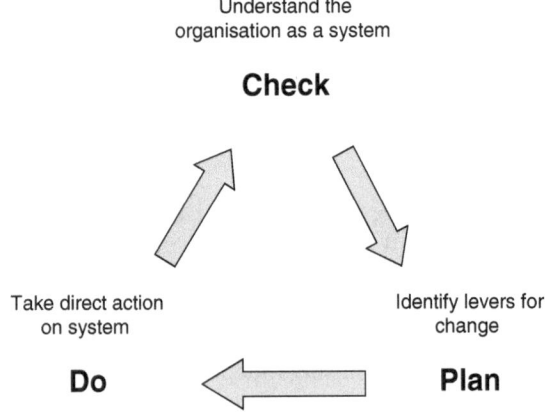

Figure 2.1 Systems thinking 'Check-Plan-Do' approach
Source: Seddon, 2005.

Table 2.1 Stages of systems thinking review

Phase	Activities	Time
Check	Study demand by collecting information and understanding what matters to customers. Define the real 'purpose' of the service from customers' viewpoints	Nov. 08 – Feb. 09
	Map the flow of work and identify waste and systems conditions that stop the flow Examine how capable the old system was in achieving the (real) purpose	
Plan	Redesign the system to deliver purpose with minimum waste	March 09 – June 09
Do	Release capacity Integrate and engrain new process, measures, and method	July 09 – ongoing

Findings from the 'Check' phase

Purpose: Figure 2.2 illustrates various stages of the 'Check' phase. The team began by defining the 'right' purpose for the service. The purpose of the system was defined from the customer's point of view as follows: 'To provide the right help for me, when I need it, to maximise my independence'. This was achieved by spending considerable time listening to demand studying 'what mattered' to customers. One team member explained that the new purpose is our *proper purpose* and what we should be delivering from a customer's viewpoint not the stated purpose (mission statement) or de facto purpose (which could be driven by targets or budget such as spending within financial year). Having examined demand on the service and after asking customers face-to-face what mattered to them the team was also able to identify the steps in the process that were of value to customers linked to this purpose. The 'value steps' were 'Get the information to find the right help for me' and 'Get the work done when I need it'.

Demand: Having defined the purpose the team continued studying the demand. Potentially only a fraction of the demand is referred for a facilities grant. Figure 2.3 illustrates that a large proportion of the demand is not referred to DFG mainly due to long lead times. Moreover, of the 68 average monthly referrals from OT service to DFG department,

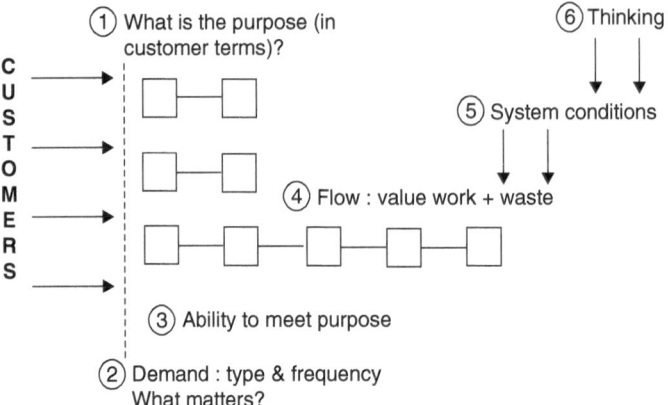

Figure 2.2 Check model of systems analysis

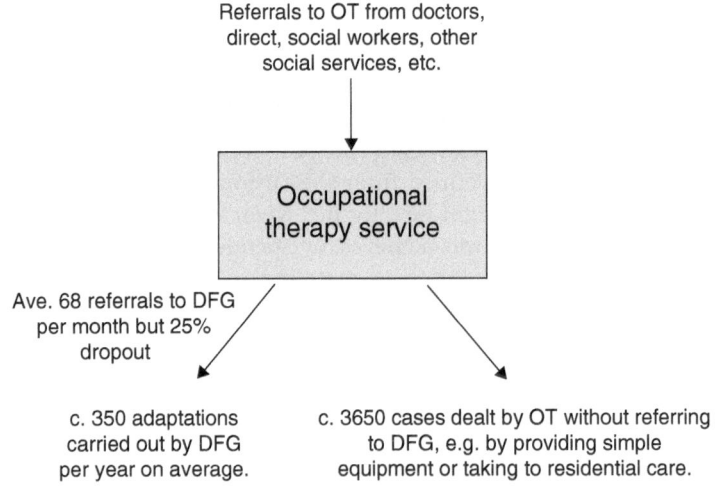

Figure 2.3 A large proportion of the demand not referred to DFG mainly due to long lead times

17 drop out due to long lead time (25 percent dropout rate). Figure 2.4 illustrates the average monthly demand over three consecutive years excluding dropouts.

The improvement team found that 92 percent of DFG's are delivered to people over 50 which account for 57 percent of Neath Port Talbot population and this figure is growing. Moreover, 56 percent of those over 50

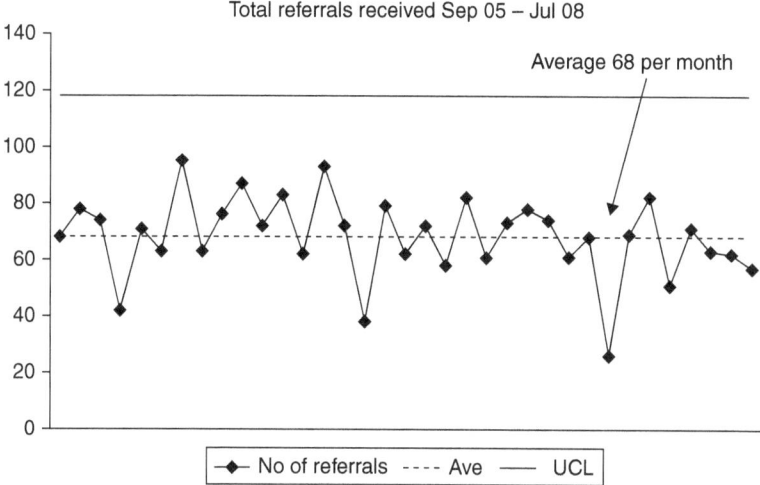

Figure 2.4 Average monthly demand for DFG over three consecutive years (excluding dropouts)

in NPT (around 30,000 people) have some sort of long term limiting illness. This evidence suggested that the DFG service had more to do with age than disability which helped put the service into context.

This was an important point of learning which led to further investigation. Further study showed that of the 750 people who went into residential care over a five year period, 244 had been identified by OT services for a DFG. Eighty-five of them had received the DFG but later went into residential care at an average age of 84. One hundred fifty-nine did not receive a DFG due to the waiting time and were admitted to residential care at an average age of 80. The evidence suggested a strong correlation between the average age of admittance to residential care and the completion of DFG works, i.e. a potential delay of four years where a DFG was received. In this case, it is suggested that four years' additional independence at home could have been possible for the 159 people who were admitted to residential care at age 80, if adaptations had been available earlier.

This period would cost on average £380 per week per person in a residential care home with a total cost of £12.57 million (159 x £380 x 52 x 4). DFG works for the same group of people, to delay admission to residential care potentially by 4 years, at an average DFG cost of £7,000 per case would have cost £1.12 million or less than 9 percent of the above figure.[1] It must be noted that this only indicates potential and hypothetical savings and that the figures are calculated in retrospect rather than actual savings.

Clearly, by providing appropriate DFG at the appropriate time people can be sustained within their own accommodation, not only reducing residential care costs but arguably delivering considerable emotional and community benefits to service users. Nevertheless, the impact of timely DFG service could go beyond delaying admission into residential care, for example to alleviate costs in the home care service or to discharge existing clients from residential care. Table 2.2 provides a few examples of individual cases referred directly to the DFG review team from social care sources. These were identified as representative cases typically managed by social and health care officers by the staff working within the service.

Table 2.2   Examples of the impact of timely DFG service

Care Type	Value of DFG Work	Value of Equivalent Care (projected)	Basis of Calculation
Residential care (can also apply to nursing care)	£18,000	£80,000	Client identified for residential care placement, delay in admission by providing shower room extension and extending bedroom via DFG.
Home care	£1,500	£12,500	Client receives commode cleaning 15 minutes daily. Offered WC installation via DFG. Based on the 2 years of service already provided and a projected further 8 years independence at home.
Discharge from residential care into independent living	£6,500	£280,000	Client currently in residential care. Suitable level access property to be sought and further adapted via DFG. Based on the 4 years of service already provided and a projected further 10 years (linked to the age of the client).
Discharge from hospital care to home	£8,000	£60,500	Client in hospital with future admission to nursing care. Level access facilities to be provided via DFG will enable independence. Based on 3 years' average wait DFG (if referral is made) and the equivalent time 'waiting' in nursing care.

The findings in Table 2.2 illustrate the importance of understanding the end-to-end system in measurement. Whereas conventional public sector targets and performance measures focus on silo-based efficiencies, activity costs, and managing the budget, the evidence suggests that the actual cost lies in the end-to-end flow of delivering the purpose of the wider system.

More important, further detailed study of the demand revealed that 71 percent of the demand presented at DFG was 'preventable' based on investigating 3,435 phone calls coming into both the OT and the Housing Renewal and Adaptation Service (HRAS). Preventable demand (or in better terms failure demand) is demand resulting from the system failing to do something or to do something right, for instance when customer asks 'I don't understand your letter?', 'why am I waiting?' or 'when will you be here?' Preventable demand needs to be addressed systematically and designed out of the system.

Simply put, attempts to improve the efficiency of the service flow without understanding the often high percentage of failure demand will lead to designing a process which is both ineffective and inefficient. That is why efficiency and effectiveness are inextricably linked in service improvement. Figure 2.5 illustrates analysis of demand by type and frequency as they were presented at the OT administrative department (1,114 phone calls were studied over 20 working days).

"Demand – type & frequency" OT admin telephone demand	
Unclassifiable – Can I speak to	506 calls
Classifiable (value & preventable)	(%)
Value demand:	38
"I am a client having problems & need help"	36
"I want to change my appointment with the OT"	2
Preventable demand:	62
"Referral made, when are you coming out - Client"	22
"Equipment enquiries – What should I do?"	5
"Incorrect call transferred"	4
"Social worker calls – when is the client being assessed"	3
Remaining Demand – Chasing calls, I don't understand your letter, Enquiries from other staff, I have a defect etc	28

Figure 2.5 Demand analysis by type and frequency

Figure 2.5 should help to further understand the concept of failure demand.

Capability to meet purpose: In the 'Check' process the ability of the system is measured against the purpose or what matters to customers. Therefore, the DFG's capability was looked at in terms of the end-to-end time measure of completing the work (meeting the value demand). As illustrated in Figure 2.6, the average lead time to complete DFG work was 675 days consisting of 435 days waiting and 240 days installation time. Again, Figure 2.6 shows that the upper control limit is 1,420 days meaning that the DFG service is predictably and reliably delivered in no less than 1,420 days. It therefore was no surprise to discover that 25 percent of those who had been referred for a DFG had dropped out from the waiting list over five years leading up to December 2008 (1,060 people). Of these:

- 234 died while waiting;
- 193 dropped out of the system due to concern about the means test;
- 115 decided to do the work themselves rather than wait;
- 45 went into residential care; and
- 271 dropped out for other reasons that are unknown.

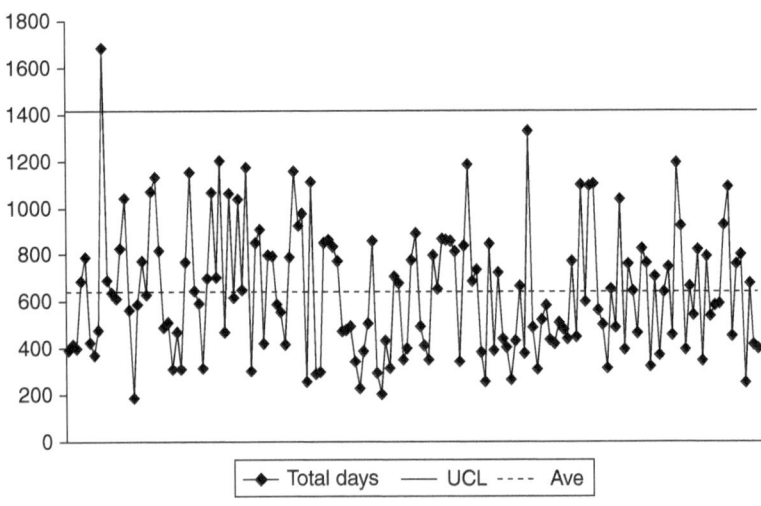

Figure 2.6 Total Elapsed Time (TET) of delivering DFG service

Moreover, the evidence showed that of the referrals made (first point of contact) in the last 12 months, 45 percent dropped out while waiting for an OT assessment. Even worse, over five years 850 people had dropped out after DFG preparatory works had begun:

- 130 died while waiting;
- 198 dropped out due to the need to pay a contribution; and
- 312 withdrew for reasons that are unknown.

The above evidence does not take into account the likelihood that many citizens were not referred for DFG in the first place because of the long lead times. The evidence suggests that by delivering an effective DFG service the average age of referral to residential care can be substantially reduced leading to a potential tenfold cost savings. Again, the evidence shows that the financial knock-on effect of not delivering service in reasonable lead times extended beyond the immediate costs to the service. That is why operational measures of system ability, such as the Total Elapsed Time, can be regarded as leading measures of performance as opposed to activity cost and budget which are lagging measures. However, the overt focus on lag measures of performance in public sector resembles driving a car while looking back in the rear view mirror.

Flow: The model for 'Check' comes to mapping the service flow only after extensive study of purpose and demand. This is essential in order to make sure that the improvement process is driven by effectiveness rather than a suboptimal emphasis on efficiency. Understanding preventable demand and being able to design the flow against real (value) demand is a basic advantage of the approach. Mapping the flow showed many hand-offs between various experts, agents, and administrators.

The end-to-end DFG process involved 291 steps, of which only 20 were of value to the customer (based on the purpose). Figure 2.7 illustrates an overview of the flow prior to redesign through systems thinking review. As shown, DFG expert staff only got involved in work after the case had been through several hand-offs while the front line staff had little expertise or authority to deal with the case. The managerial thinking behind such system design, often, is that expert staff are too expensive and too busy to deal with demand at the first point of contact.

System conditions and management thinking: As illustrated in Figure 2.2, in order to redesign an efficient service against the purpose,

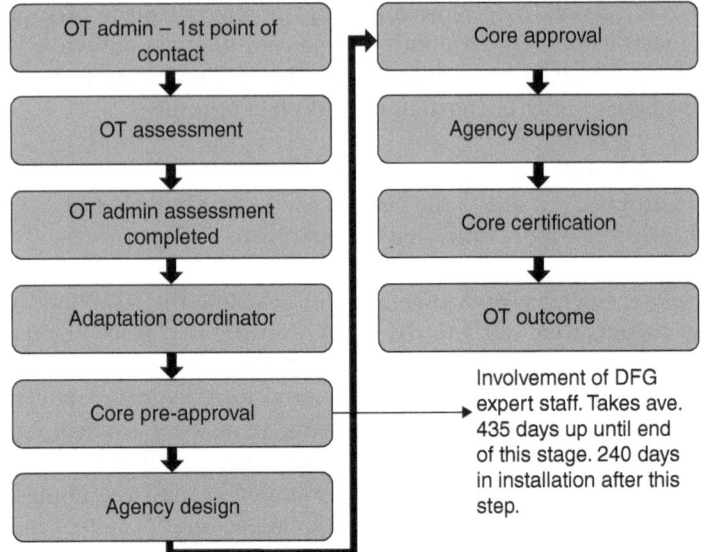

Figure 2.7   Flow of DFG process before redesign

it is important to understand the system conditions which drive the current performance as well as the managerial assumptions which inevitably underpin the system design. System conditions in DFG were exposed by asking front line staff what got in the way and prevented them from doing a good job. They replied targets, procedures, controls, authorisation requirements, and IT systems. These system conditions could often become de facto purposes for the system. The national performance indicator for delivery of a DFG is a good example of how targets or centrally imposed measures become a de facto purpose and drive the wrong behaviour.

The national indicator measures the time taken to deliver a DFG from the first point of contact to the completion of work end-to-end. The indicator is in effect a 'league table' comparator across Wales which, maybe unintentionally, sets a competitive target for authorities. It presumes that all authorities follow the same process and are working on the same end-to-end times. However, in practice, DFG delivery process across Wales is not consistent and the measurement of the data is not necessarily a true reflection of performance or a true measure of efficiency. The DFG service could fall within various parts of Local Authorities and the lead-time could vary depending on the stop/start of the clock. This arguably leads to the design of DFG systems to achieve the de facto purposes aimed at improving the ranking rather than providing the best solutions for disabled people. The indicator offers only one piece of 'evidence', namely the average time taken by each council to deliver a DFG. For example:

- it fails to capture differing levels of demand received at authorities (in certain areas like Neath Port-Talbot the disabled population is much higher than that national average);
- it does not indicate the variation between Councils in DFG delivery processes especially with regards to the way DFG operates within the wider system;
- it does not show the different levels of priority placed on DFGs by each Authority; and
- it does not indicate the financial commitment of different Local Authorities.

The Welsh Assembly Government's annual published results place Neath Port Talbot with the 'poorest' performing authorities. The continued use of the indicator may be damaging to DFG services across Wales since it does not compare like with like but affects staff morale and could lead to higher costs where improving a Council's ranking becomes the purpose, rather than delivering DFG services that help the customer. A typical example of its effect is outlined in Figure 2.8, where

Example: *Mr & Mrs A (88 and 84 years of age respectively) required a ground floor shower facility due to Mr A having bone cancer. Mr A slept on the ground floor as a result. Mrs A has other disabilities but is also Mr. A's carer. While preparation for delivery of the DFG was underway Mr A passed away. The works were also needed for Mrs A. Due to the 'de facto' purpose linked to the national indicator and the council's budget arrangements; Mrs A was pressed to allow works to start in January 2009, to ensure completion by the end of March. This involved work progressing within three months of her husband passing away and at the coldest and wettest time of the year. Mrs A continued to occupy the house while the demolition works were carried out to provide access to the new shower room. The period of the construction led to increased costs due to foundation problems caused by the wet weather, and clearly caused Mrs A distress and discomfort.*

Figure 2.8   An example of how systems conditions drive wrong behaviour

in order for the DFG works to hit indicator timescale 'targets' at the end of each financial year, DFG related installations are started in January to ensure completion before April.

The following chart shows DFG expenditure over 6 years; the budget and national performance targets visibly drive the whole DFG system leading to huge spikes towards the end of each financial year marked in red. In order to experiment with redesign, the system conditions were removed in the DFG service at NPT. This sent an essential message to staff that management thinking had changed and showed senior leaders' commitment to change. Figure 2.9 explains the redesign phases of systems thinking review, i.e. 'Plan' and 'Do' stages.

Plan: Following the completion of 'Check', the 'Plan' phase involved experimenting to find a better system which, in achieving the purpose from the customer's point of view, is also simpler and cheaper. Given the data collected during 'Check', this also involved looking at the cost benefits for social care and health care that can be found by early DFG intervention. Briefly, the aim of the 'Plan' was to redesign the DFG service to:

- respond to demand, based on the new 'purpose';
- achieve 'perfect flow';

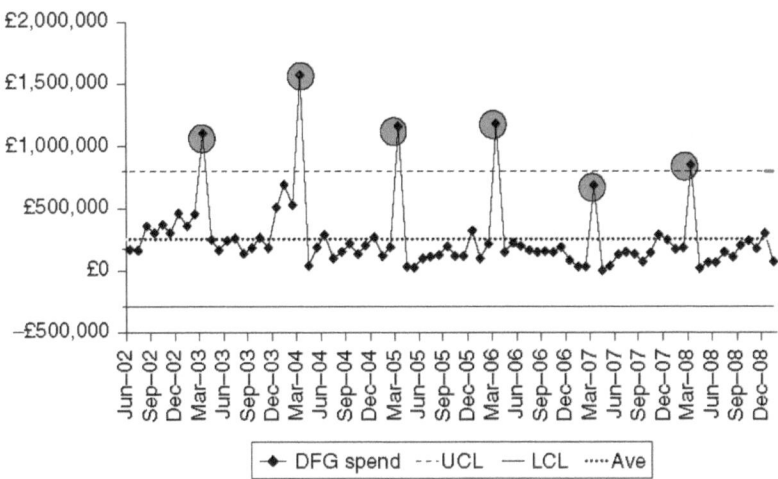

Figure 2.9   DFG expenditure pattern and target driven behaviour

- develop measures to help in understanding the work and enable improvement (understanding and acting on variation is an integral part of measurement); and
- ensure management roles added value to the new system.

The above analysis revealed that performance is driven by system conditions which in turn are rooted in management thinking. Therefore, in order to change the system it is necessary to change the underpinning management beliefs. The review team decided that the new DFGs should be provided based on the following 'systems thinking' management principles as designed by the senior leaders directly involved in the review:

1. Redesigned services should respond to demand, based on the 'purpose'.
2. Managers check and act upon the system.
3. Employees are motivated to do a good job.
4. Those working in the system should design the system.
5. Decisions must be based on data and evidence.
6. Managers facilitate process of change and redesign to meet purpose.
7. Managers should create a supportive environment to encourage individual decision making to deliver the purpose.
8. When employees raise an issue, managers must jointly visit the problem then act.
9. Managers should regularly spend time walking the system end-to-end.
10. The system should provide help to the customer going through the process.
11. IT should support the whole system end-to-end.
12. Use measures to tell us how we are achieving the purpose.

The above principles show a radical shift away from old ways of management thinking. The leader of the service said 'if we focus on delivering service to citizens in the process of doing so we also become more efficient'. Thus the redesigned system was driven by a profound understanding of the way work worked. Based on the above principles the team asked themselves 'how would you deliver service from a blank sheet of paper?'

The improvement team experimented with the redesign to ensure it was delivering results against the right measures. Therefore the 'Check' team also became responsible for implementing the redesign. One of the key aims of the redesign was to eliminate non-value adding steps where possible. For example, expert staff were brought to the front allowing them to face the customer at the first point of contact. This meant that the redesign team met customers when they required them (which suits customers), rather than handing off work to others (unless it is essential) and they arrange for installations to start at a time which suits the customer.

Staff were closely involved in the examination of the effectiveness of the old system to ensure that front line knowledge drove the redesign. For those involved in the improvement, the experience has been dramatic. Officers have commented that the review is the most important piece of work that they have ever been involved in. One of the managers involved in redesign commented to the author 'if we had been through redesign without 'Check' then it could have been very different'. Discussions with the redesign team revealed that the action learning approach has had a drastic effect on their ability to redesign. Also, the team commented that the knowledge-based approach (i.e. decisions based on facts only) was very different from using experience which could be influenced by biases and assumptions.

Figure 2.10 shows the flow in the redesign system. In this model the officers get on site as quickly as possible (subject to when the customer wishes to be paid a visit by the officers). More important, various holders of expertise within the team (i.e. occupational therapist, surveyor, adaptation coordinator, and even installation contractor) were present for the first visit and work was dealt with there and then without any hand-offs. A fundamental problem with hand-offs is that often information passed between experts is not 'clean' (for example, inaccurate or incomplete) leading to 'failure' demand.

Moreover, the team was able to glean all required information from the customer and eliminate yet another major source of 'failure' demand. Furthermore, the redesign team might provide additional help such as benefits assessment, essential repairs, minor equipment, energy advice, and health and safety advice, all of which aimed to keep people living longer in their own home. The DFG redesign model intervenes as early as possible because data showed that early response reduced costs and further calls on social services.

During this phase 15 cases were used as an experiment to test the redesign. Another key feature of redesign is that the same surveyor takes the build process (design, cost, supervise, and certify) through from start to finish. Also the team started collecting data (based on more meaningful measures) to support continuous improvement. Early

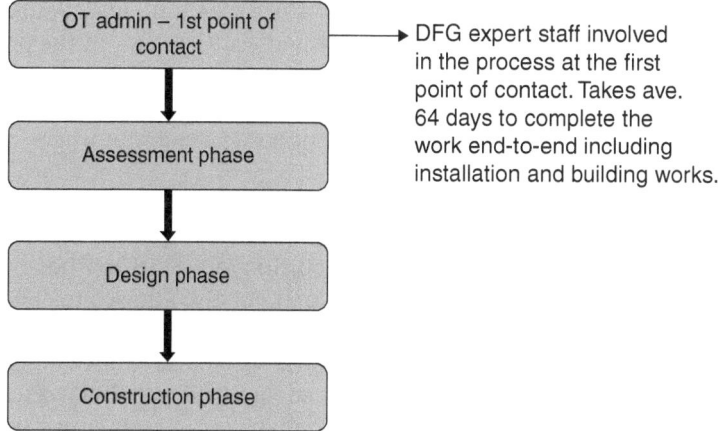

Figure 2.10   Flow of DFG process after redesign

evidence on the benefits of the redesign was positive. As a result of reducing lead times and creating space for officers to think, the liaison officer on the team had the opportunity to investigate consider alternative procurement processes such as materials supply, specification, and contractor options. This resulted in finding options to reduce the cost of installations from £7,000 down to around £6,300. A Senior Surveyor on the review team commented that:

> Savings have been achieved by getting the work right first time. This is a result of the improved communication between the customer, OT, surveyor and contractor. 'Check' showed an average £1,200 of variation on each grant. As a result of identifying this data 160 cases were reviewed in detail. 25 per cent of the variation was found to be a result of re-work due to failure to get the work right first time. Eliminating this will reduce the overall cost of the grant. Also through encouraging experimentation and continuous improvement further savings have been identified, for example by negotiating discounts with suppliers and investigating alternative arrangements for tendering the supply and installation of specialist equipment.

Do: This report was written during the early stages of the 'Do' phase and therefore full explanation of the outcomes is not yet possible. However, the redesign team is disseminating the redesign by gradually involving the entire DFG team (around 30 officers) in the redesign way of working. This means that the other officers become fully aware of 'why' the system is redesigned this way and get 'their say' in the process. One of the key principles at this stage was to have a planned roll-in to take staff from the old to the new system. In contrast to most change programmes, this has been the opposite of roll-out where staff are expected to follow new methods.

In order to achieve this, other teams are being formed with one expert from different areas in each team (for example, occupational therapist, surveyor, and adaptation coordinator). Staff are not only trained in the 'systems thinking' approach but also allowed to experiment against purpose and demand to create maximum buy-in and achieve the same 'action learning' and profound understanding through experimentation. This aims to produce sustained results. Another key aim of the 'Do' phase is to make sure the right measures are implemented and that the system is capable of monitoring and reacting to demand by means of the right measures. The following explains more about measures.

Results before and after

The following measures are used to gain knowledge and understanding of the system, and review results of experiments:

- end-to-end times;
- variations (e.g. additional costs);
- delivering service right first time;
- demand (e.g. volume in);
- cost of works (for example, work categories, contractor performance);
- staff morale (e.g. sickness);
- expenditure (e.g. monthly); and
- client satisfaction based on two questions:
  - 'Please rate the service given on a scale of 1 to 10 (10 being highest)'; and
  - 'If the service is not rated as 10, why not?'

According to the redesign team and service leaders these measures may be turned on and off as required by service officers or could be replaced by new measures if necessary. Each measure can be broken down from a full service picture to reflect the progress of teams, individual officers, work categories, specific projects, or ongoing experiments. Table 2.3

Table 2.3 Key improvements to DFG service

DFG measure	Old system	Redesigned system*	Comments
Average. end-to-end time	675 days	64 day	675 made up of 435 wait and 240 install
Flow steps (end-to-end)	291 steps	34 steps	Every step from first point of contact to completion of works
Preventable demand	71 percent	40 percent	See above
Costs of delivery (Average per grant)	£499	£319	Staffing activity costs (36 percent improvement)
Cost of physical works; average per case	£7000	£6300	Procurement savings & reduction in re-work
DFG drop outs	33 percent of cases	Nil	Early intervention prevents dropouts

* The results are based on a sample of 39 completed cases. The redesign of the service is still evolving and continuous improvement may lead to further improvements.

illustrates some of the key measures and results from the 'systems thinking' improvement.

Conclusions

The leaders and team members equally stated that the change process could be difficult at the beginning, especially for managers who set up the process in the first place. People's familiarity with old ways of working can cause a difficult spell at the beginning. They also explained that the emphasis on facts (scientific approach) and designing improvements as action learning (learning by being involved at the front line) instigated a profound realisation within the team. This can be assessed as an important feature of the systems thinking approach. In this case senior leaders and DFG officers alike sat down to listen through calls or went out on visits with occupational therapists. The improvement team felt that now 'management can better understand the problems we have'.

Another interesting finding is the level of engagement of the DFG officers or the 'doers'. Public sector managers and central government are often faced with the dilemma that implementing new and more efficient systems could lead to losing employees' engagement in new ways of working. Some organisations attempt to communicate or to 'explain' the need for the new system by sending middle managers on training programmes. However, this hardly solves the problem since explanation is barely synonymous with understanding which comes from a far deeper and more hands-on engagement with the work itself. At the same time, it is no wonder that change does not always stick if workers do not appreciate the necessity of the new way of working. Organisations and managers, who think they see the benefits of the new system, often respond by forcing change through 'with an iron fist'. The result is the creation of 'dumbed-down' systems within which the role of the worker is reduced to a powerless doer. As such, systemisation by a central authority simply eliminates 'thinking' from working and improvement. This is clearly in stark contrast with the very core of continuous improvement and systems thinking which aim to put thinking back into work, back into the frontline.

The team members in this case expressed concern regarding 'best practice' or 'guidelines' from the centre since they often 'don't make sense' and lack contextual knowledge about the way work works. One member of the redesign team said, 'what we find is that best practice tends to be good pieces from different systems put together and imposed on all different authorities which doesn't necessarily lead to better performance'.

Furthermore, the improvement has intensely engaged various layers of the organisation due to an emphasis on system conditions and underlying thinking that drive the existing behaviour of the system. It is widely accepted in the systems thinking literature that system structure and system conditions drive behaviour. It is also established that in order to change behaviour, it is important to change the thinking that underpins the system structure and conditions.

The lead officer for the DFG service commented that 'as with most of the service staff, front line mangers and senior leaders I believed that the system was appropriate for the needs of service users and I was trying to do a good job. However, I came to understand that the controls, targets, procedures and service standards which were put into place to "manage" the service, had in fact led to the separation of the decision making from the work leading to the introduction of waste, demotivation of front line staff and increased delivery cost. Unfortunately well intentioned policies had led to wrong results. Acceptance of this has changed my thinking and my understanding of how the service should be delivered. This is an experience that I believe every public service manager would benefit from'.

The author's observation is that 'dumbed–down systemisation' is ubiquitous in the public sector, from local authorities to health. By putting expert staff at the frontline who are enabled to think and absorb the requirements of the customers, this local authority has not only delivered savings but also improved service waiting times considerably. Interestingly, the activity cost of service delivering DFG has only decreased by 36 percent (from £499 down to £319) whereas the wider systemic costs could decrease at an even higher rate as shown in the operational measures in Table 2.3. For instance, independence at own residence by potentially (on average) four years across the sampled cases, £700 savings in the average grant size and reduction of the service delivery from 675 days to 64 days (90 percent improvement).

The renowned systems thinker, Peter Senge (1990; p. 53) says, 'the deepest insight comes when they [managers] realize that their problems and their hopes for improvement are inextricably tied to how they think. Generative learning cannot be sustained in an organisation where event thinking predominates. It requires a conceptual framework of structural or systemic thinking, the ability to discover structural causes of behaviour'. Senge credits Chris Argyris for this realisation. Distinguishing between single and double-loop learning, Argyris (1999) explains that: 'Single-loop learning occurs when matches are created, or when mismatches are corrected by changing actions. Double-loop learning occurs when mismatches are corrected by first examining and altering the governing variables and then the actions'.

For Argyris and Schön (1974), single-loop learning involves improving incrementally through learning new skills and capabilities, doing something better without challenging the underlying beliefs and assumptions. Double-loop learning goes further than single-loop learning by reshaping the patterns of thinking and behaviour which govern why actions are taken (see Figure 2.11).

Argyris and Schön credit Ashby's (1952) use of the metaphor of a household thermostat to give an example. When the temperature oscillates around a single point with the heating unit turning on and off in reaction to temperature changes, this could be characterised as single-loop learning: the feedback system reacts in a straightforward way to fluctuations in the environment. When the householder intervenes and alters the thermostat, this can be seen as the equivalent of double-loop learning: the human feedback loop here connects the household temperature not only with the heating unit, but with the thermostat

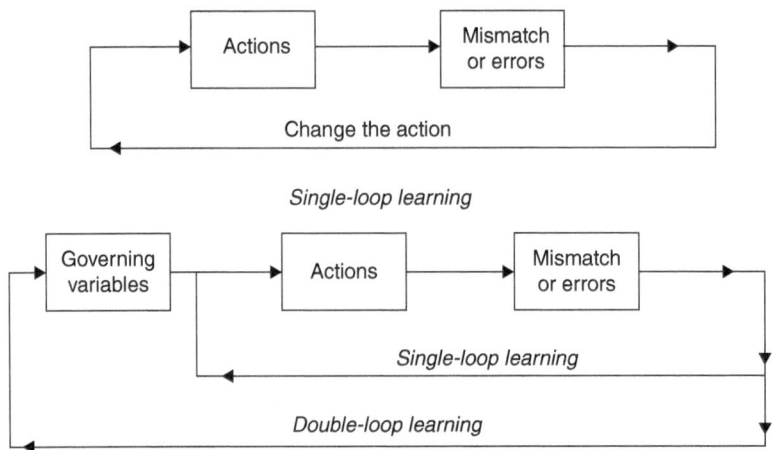

Figure 2.11   Single- and double-loop learning
Source: Argyris 1990 p. 94.

settings around which the temperature will oscillate. 'Double-loop learning changes the governing variables (the "settings") of one's programs and causes ripples of change to fan out over one's whole system of theories-in-use' (Argyris and Schön 1974; p. 19).

By recognising that there is a system of interaction which underpins one's actions, it is possible to change and in the process become more open and self-aware. Seddon (2005) describes this as the need to 'unlearn' before one can 'learn' the new way that a system should work, in an emergent and adaptive approach to change. 'Systems thinking is only truly learned by doing, by action learning: it is only by doing that managers can unlearn, can find out for themselves where their current beliefs about the design and management of work are flawed, in order to put into place something that works systematically better, and can systemically be further improved' (Seddon and Caulkin 2007; p. 15).

Embracing the emergent approach to change is positioned firmly in the systems thinking tradition: 'The idea of emergent properties is the single most fundamental systems idea and to use this (and other) systems ideas in a conscious organised way is to do some systems thinking' (Checkland 1997, cited in Chapman 2002; p.38).

Chin and Benne (1969) called this form of action learning 'normative re-educative', whereby people are taken through their work to surface their underlying non-conscious values and then to change them, using what they call 'temporary systems' as a medium of re-education. Argyris

(1999, p. 90) suggests that 'it is possible to help individuals learn new theories-in-use and to create new learning systems. The intervention requires the creation of a dialectical learning process where the participants can continually compare their theories-in-use, and the learning system in which they are embedded, with alternative models. This requires that interventionists make available alternative models with significantly different governing values and behavioural strategies'.

In the above case study a normative re-educative intervention took place which helped staff contrasts command-and-control against systems thinking, enabled people to test their theories-in-use against the real purpose, performance, and requirements of the service.

Note

1. It must be noted that the cost of not going to residential care is not just the adaptation (i.e. £7000 towards DFG) but also any home care package received by the client. However, analysis showed that only 2 percent of those who received DFG's later received home care. A more detailed analysis should also take into account the average cost of home care. Moreover, arguably the total cost of the client in residential care is less than £380 due to means testing and client contribution. Unfortunately, no reliable average existed. Therefore the conservative estimate of £380 was used for the analysis which was the minimum in a report covering 800 cases who received residential care. Further research is recommended to establish a reliable average.

References

Argyris, C. (1990) *Overcoming Organizational Defences: Facilitating Organizational Learning.* Allyn and Bacon: Boston, MA.

Argyris, C. (1999) *On Organisational Learning.* Blackwell Publishing: London.

Argyris, C. and Schön, D. (1974) *Theory in Practice: Increasing Professional Effectiveness.* San Francisco: Jossey-Bass.

Ashby, R. (1956) *An Introduction to Cybernetics.* Chapman and Hall: London.

Seddon J. (2005) *Freedom from Command and Control.* Vanguard Press: Buckingham.

Seddon J. and Caulkin S. (2007) 'Systems Thinking, Lean Production and Action Learning' in Action Learning Research and Practice, 4 (1), April 2007, special issue: 'Lean Thinking and Action Learning'.

Senge, P. (1990) *The Fifth Discipline.* Random House Business Books: London.

3
Systems Thinking in Adult Social Care: How Focusing on a Customer's Purpose Leads to Better Services for the Vulnerable in Society and Enhances Efficiency

Brendan O'Donovan

This chapter documents the application of systems thinking methods to an English local authority Adult Social Care department. The author shows how the perception of large demand for scarce social care resources leads councils to screen out many of the applicants for this service through the strict application of eligibility criteria. A systems analysis shows the inefficiency of this approach: many of the users are later found to require a more expensive service once their condition has deteriorated sufficiently to be eligible. By redefining the purpose of this service and refocusing on doing what matters to the end user and ensuring it is done right-first-time, the social workers in the system are enabled to experiment with new methods of providing the service. The results from before and after the experiment are then examined, showing both cost savings and improved operational measures for service delivery.

Introduction and Methodology

This case study examines the application of systems thinking in an English local authority Adult Social Care (ASC) department. Systems thinking in the form of the Vanguard Method has been documented as applied to housing (ODPM 2005; Jackson, Johnstone and Seddon, 2007; McQuade, 2008) and other public services such as policing and housing benefit payments (Seddon, 2008). Here the aim of the study is to understand and extend experience about a contingent, contextual, and complex process, i.e. the application of systems thinking improvements in a social care setting. Evidence of performance in this case includes operational data from the intervention including:

- the end-to-end times for cases
- the number of 'first-time fixes'
- costs per case

Evidence also included interview descriptions by workers of comparisons between the old and new designs. In this council, interviews were conducted with participants at different levels in the organisation as well as with the facilitator who had been leading the systems thinking interventions. The interviews were semi-structured, non-standard respondent interviews where the purpose was to gather data which was reliable, valid and relevant to the research questions and objectives. These interviews were recorded, listened to again and transcriptions made of pertinent sections to the research, as suggested by White et al (2009).

Hypothesis

The 'Check-Plan-Do' systems thinking methodology was devised by Seddon and documented in detail in his books (Seddon, 2005 and 2008). In these accounts of how to go about systems thinking interventions, Seddon explains that Taguchi's Quality Loss Function (Lochner and Matar, 1990), with the associated concept of setting a nominal value and trying to improve against it through experimentation rather than working to arbitrary tolerances, should be adapted and applied to services (Seddon 2005 p. 60). Simply put, 'in service organisations it is the customer who sets the nominal value' (Seddon, 2008 p. 69). The research hypothesis of this paper is that delivering to the customer's nominal value right-first-time would show a decrease in the organisation's costs, as one element of the reduction in the 'loss to society' encompassed within Taguchi's Quality Loss Function.

Whilst many operations management and marketing writers such as Porter (1985), Womack and Jones (1996) and Vargo and Lusch (2004) have mentioned or even offered definitions of value to the customer, it can be argued that none has developed an effective 'operational definition' (Deming, 1982 p. 276; Neave, 1990 p. 110) by which managers could identify and manage value in their operations. In services, systems need to be designed in such a way as to be able to provide what the early systems thinker Ashby called 'requisite variety' (Ashby 1956) in order to satisfy the customer's 'nominal value'.

Through this case study in an Adult Social Care (ASC) service, it will be necessary to show the underlying importance of designing the system in such a way as to detect the true nominal value of a customer and then ensuring that a service can deliver against it. By addressing this subject, this case study adds to the body of knowledge on customer value in the particular context of an ASC system. Moreover, this chapter explains how beginning with a focus on the customer's nominal value and thus putting effectiveness measures before arbitrary measures of efficiency leads to both better service and lower costs.

Policy background to adult social care and this intervention

Adult Social Care (ASC) consists of services aimed at helping people to live as independently as possible. More than 1.5 million people use social care services in England. Whilst the largest group of users for publicly funded social care are the over 65s, other groups of people in receipt of social care can include those with sensory impairments, physical or learning disabilities, terminal or mental illness, problems connected to ageing, alcohol or drug dependencies (CQC website, 2009). Types of ASC services can include home care, day care, residential care, meals in the community (commonly known as meals on wheels), specialised advice and support for people with hearing, sight, or speech problems, car 'blue badges' for people with a permanent disability, equipment, adaptations and telecare alarm systems to help people live independently at home and other help for carers. There is a distinction drawn in England and Wales between acute healthcare which takes place within the remit of the National Health Service (and is thus free to the user) and the ongoing care of a patient which falls within the remit of local authority ASC departments (meaning that users may be subjected to means testing for the service). In practice, this distinction between which type of service is required by the end user can be difficult to make and can cause delays in provision as local authorities and the NHS argue over who is responsible for a patients' costs.

Screening out service users

ASC services in England are nominally open to any citizen requiring assistance. However, as concerns have grown over both a perceived scarcity of resources and 'postcode lotteries' (where local services varied greatly across different geographies), national guidance has been published to advise who should be eligible to receive services. This guidance was laid out in the Department of Health's 2002 circular to local authorities 'Fair Access to Care Services (FACS) – guidance on eligibility criteria for ASC' (DH, 2002). The FACS criteria are therefore meant to guide all local authorities who have to decide where to pitch their thresholds for access to services within a framework. The four bands within which people's eligibility needs are separated by FACS are 'critical, substantial, moderate, or low'. Essentially, these criteria are now used by authorities to try and screen out what is perceived to be more demand than they have the resources to cope with. Table 3.1 shows when someone is judged to be categorised as one level or another.

'The case for change' (Department of Health, 2008)

On top of the eligibility criteria and other regulations for social care provision, there are many discussions about how ASC will be provided in the future. There is a fear that with ongoing demographic change (an ageing but longer living population), demand for services will grow. With the belief that there will be more people competing for increasingly scarce resources, the government has called for the ASC system to be 'transformed' in line with other UK government 'public sector reform' initiatives, e.g. 'Transformational Government' (Cabinet Office Strategy, 2005). A government consultation exercise was undertaken in 2008 which discussed the need for new, more 'personalised' services (HM Govt 'The Case for Change' May 2008 p. 9). It is against this policy background that local authorities have been searching for new ways to manage their services. In order to attempt to reconcile these ideals of 'person-centred' services with the perceived need for rationing of resources, the local authority involved in this study decided to undergo a systems thinking intervention.

Systems thinking in ASC: The story of 'check'

This local authority decided to apply systems thinking (Seddon 2003, ODPM 2005) in their ASC department. The stated purpose of the intervention was to:

- Reshape and improve the customer journey through Assessment and Care Management; and

Table 3.1   The FACS band classifications

Critical – when	Substantial – when
●life is, or will be, threatened; and/or ●significant health problems have developed or will develop; and/or ●there is, or will be, little or no choice and control over vital aspects of the immediate environment; and/or ●serious abuse or neglect has occurred or will occur; and/or ●there is, or will be, an inability to carry out vital personal care or domestic routines; and/or ●vital involvement in work, education or learning or will not be sustained; and/or ●vital social support systems and relationships cannot or will not be sustained; and/or ●vital family and other social roles and responsibilities cannot or will not be undertaken.	●there is, or will be, only partial choice and control over the immediate environment; and/or ●abuse or neglect has occurred or will occur; and/or ●there is, or will be, an inability to carry out the majority of personal care or domestic routines; and/or ●involvement of many aspects of work, education or learning cannot or will not be sustained; and/or ●the majority of social support systems and relationships cannot or will not be sustained; and/or ●the majority of family and other social roles and responsibilities cannot or will not be undertaken.
Moderate – when	**Low – when**
●there is, or will be, an inability to carry out several personal care or domestic routines; and/or involvement in several aspects of work, education or learning cannot or will not be sustained; and/or ●several social support systems and relationships cannot or will not be sustained; and or ●several family and other social roles and responsibilities cannot or will not be undertaken.	●there is, or will be, an inability to carry out one or two personal care or domestic routines; and/or ●involvement in one or two aspects of work, education or learning cannot or will not be sustained; and/or ●one or two social support systems and relationships cannot or will not be sustained; and/or ●one or two family and other social roles and responsibilities cannot or will not be undertaken.

Source: Department of Health 2002; CSCI 2008.

- Release capacity from the existing system to allow for further improvement in and reshaping of the service and to help meet financial pressures from issues like the anticipated demographic growth.

The intervention followed the 'Check-Plan-Do' cycle beginning by studying the ASC work as a system (see Figure 3.1).

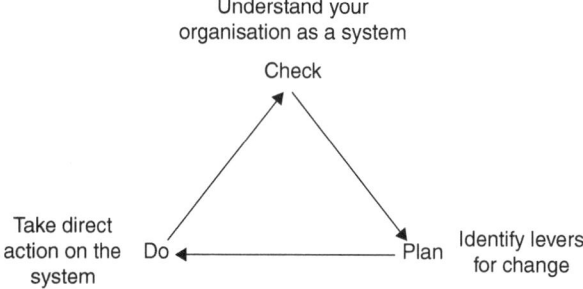

Figure 3.1 'Check-Plan-Do' cycle
Source: Seddon 2005.

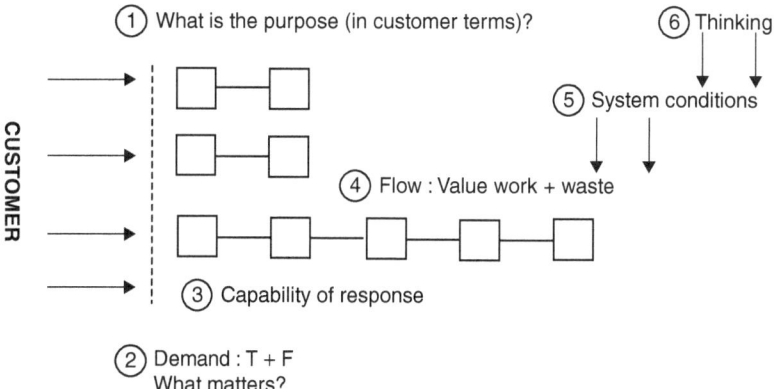

Figure 3.2 The model for Check
Source: Seddon 2005.

The model for 'Check' gives a structured way of understanding transactional services from the customer's point of view and is defined as an analysis of the what and why of the current system (Jackson et al, 2007).

The process of going through Check 'helps to identify scope for improvement in the current system that had previously been hidden' (McQuade 2008 p. 57). In order to study Check, a team of frontline workers was assembled consisting of practitioners from within ASC. In parallel, operational managers and the leaders of the organisation were engaged in similar activities.

Findings from Check: Purpose of the ASC system

After some initial time listening to the demands for assistance that were coming into the service, the Check team were able to agree on a definition for the purpose of their service. The purpose of ASC from the customer's perspective was agreed to be:

- Help me live my life the way I want to

This contrasted with the de facto purpose of the current system which was:

- Help me live the life you want me to, as dictated by government policy (or our interpretation of it)

The participants within the system found it easy to define what the purpose of the system was, in simple terms, and from the customer's point of view.

The concept of nominal value

The application of Taguchi's empirical approach to defining a nominal value (instead of working within tolerances) has been adapted by Seddon for services. In the 1950s and 1960s, Taguchi 'developed a comprehensive approach to quality which touches every aspect of a product's design, manufacture and use' (Lochner and Matar, 1990 p. 5). Taguchi believed it would be better for a manufacturing manager to set a nominal value and to encourage his team to work to continually reduce variation, resulting in better product quality and lower costs. In order to capture this concept graphically, Taguchi proposed the use of a 'Quality Loss Function':

> The value of the quality-characteristic is registered on the horizontal axis, and the vertical axis shows the 'loss' or 'harm' or 'seriousness' attributable to the values of the quality characteristic. This loss is taken to be zero when the quality-characteristic achieves its nominal value, but is positive otherwise. However [...] very little loss is incurred while the quality-characteristic is fairly close to the nominal value. But, as the value moves away from the nominal, the loss increases at an ever faster rate (Neave, 1990 pp. 173–4).

As Neave says, the loss function 'keeps in our minds the necessity for continual improvement – if there are discrepancies from nominal (and there always will be), then loss is being incurred, so the need for improvement (reduced variability) is ever-present' (Neave, 1990 p. 175). In service industries, Seddon argues that it is the customer who sets the nominal value (Seddon, 2005 p. 60).[1] If an organisation does not recognise and respond to what matters to the customer, then the service experience is poorer and the organisation is forced to consume extra resources to resolve the situation.

There are two elements of the nominal value to customers in services which are uncovered in the Check process: the underlying demand being presented to the system (what service is it that the customer wants?) and also what mattered to their customers (how does the customer wish to receive that service?).

Demand and 'What matters'

By studying the demand at all the points of contact with this system (such as the demands that came into the call centre, referrals from the hospitals and other professionals, etc.), the team was able to gather knowledge about the type and frequency of requests that they were dealing with. Seddon separates value and failure demands: value demands are the demands an organisation exists to serve, the demands for the things a customer wants. Failure demand is 'demand caused by a failure to do something or do something right for the customer' (Seddon 2005 p. 26). It was found that overall, the value demands only amounted to 26 percent of the overall demand, whereas the level of failure demand was running at 74 percent.

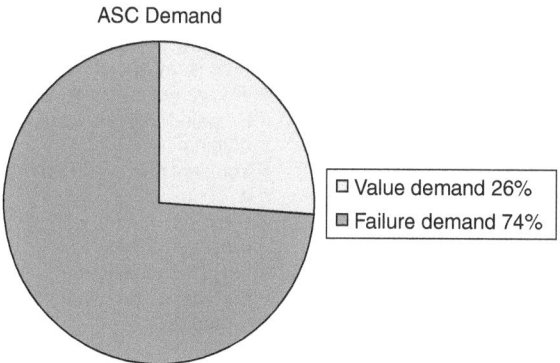

Figure 3.3 Demand into the ASC system

To give an example of the types of value and failure demands, some of the most frequently heard are listed below.

The team also identified that 80 percent of the demand entering the system was from or concerning service users that had already had some form of contact previously with the organisation. It was a surprise for the 'Check' team to discover this information and to realise that the system was currently treating all demands as if they were brand new to the system. This finding about the nature of demand challenged a current management (and governmental policy) assumption that a high degree of contact was from new service users.

Knowledge of what demand is predictable gives the first piece of knowledge that defines the customer's nominal value, i.e. what the customer wants. The second part is to answer 'what matters to the customers about how they receive their service?' In the process of analysing demand, it was possible for the team also to discern the 'what mattered' behind many of the requests, which they aggregated into two underlying concerns:

- Please solve my problem and get it right first time
- Please do it as soon as possible

From this understanding of 'what mattered' to service users, it was then possible to translate these ideas into new measures for the system.

Value

- Can I make a new referral?
- My circumstances have changed
- Can I apply for/renew my ...?
- Can I have ......(a piece of equipment)?
- Can you advise me please?
- We need help
- I want to cancel my ...

Failure

- Where is my carer/meals on wheels?
- I am know to you but my social worker has changed/left
- Who is my social worker?
- We are not sure what is going on
- I called about...but have heard nothing
- You sent me ... but I do not understand it
- I am not happy with what you have provided/offered
- I'm just checking that...
- My meals on wheels food has not turned up

Figure 3.4   The top value and failure demands

The use of the new measures allowed the team to retrospectively establish how good it had been at delivering services in the manner important to customers. In this system, these new measures were:

- End-to-end time – defined as the time taken from the point the service user first presents in the system and asks for help until the team have achieved their purpose and delivered against all the user's needs.
- Right-first-time – determined by whether the team have been able to provide the correct service(s) at the right time for the service user at the first time of asking.

It was also apparent to the Check team, as social workers witnessing the collection of data by their colleagues, that the way in which the information about the customer's nominal value was collected was very important. This insight fed into the team's 'operational principles' for redesign, discussed below.

Capability

The next step in the 'Check' process was to examine the system's capability to deliver to customers. To do this, the team analysed a cross-section of cases over 506 representative (i.e. including a mixture of different demand cases taken historically from the original point of contact ('I need help') and tracked them to the point where the recipient could say 'I can live the life I want to' so that they could understand the capability of the current system to deal with demand. The requests for Meals on Wheels, Day Centres, Occupational Therapists for adaptations, etc. were taken at random from different area offices over a period stretching back three years.

Results from Check included that:

- For end-to-end times for all demand, it was discovered that a service user would wait an average of 138 days (4.5 months) to get a service
- In 16 percent of cases, even then the system was unable to satisfy the needs of users
- Users could wait up to 486 days (15.5 months) to get a service (this is the upper control limit in the following diagram)

This information was shown in a Statistical Process Chart (SPC) (as shown in Figure 3.5) and clearly showed an upward shift in end-to-end times.

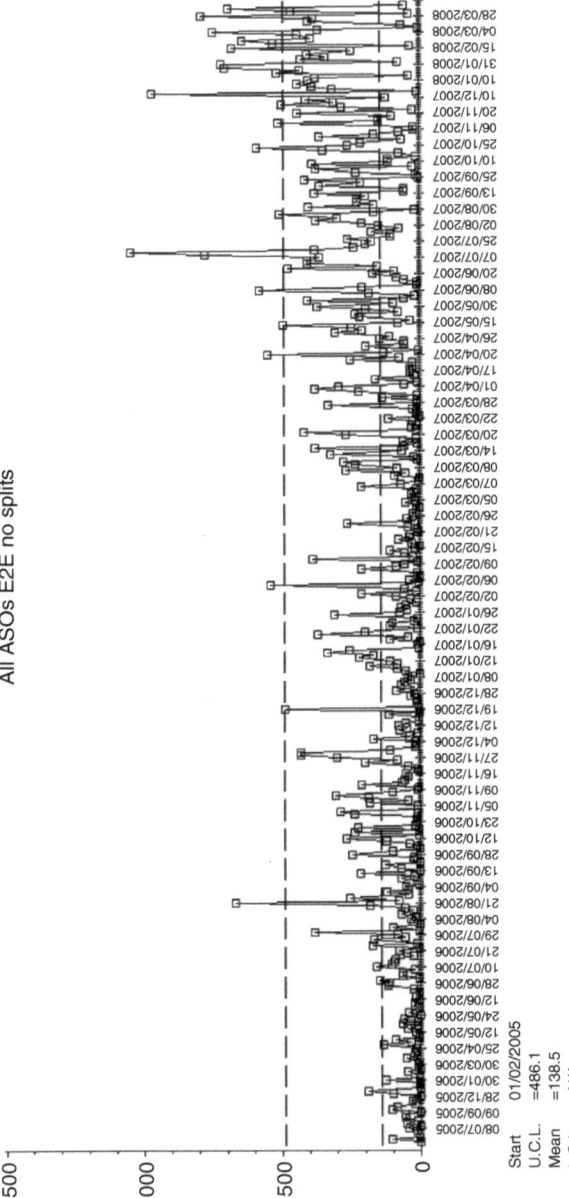

Figure 3.5 Statistical process chart showing a cross-section of ASC cases and how long they took to be completed from the customer's perspective

A split of the data based on these upward trends indicated that as controls had been tightened in response to budgetary pressures, end-to-end times for customers in the last year of the data show an average of 253 days to get services and a service user could wait anywhere up to 760 days (the upper control limit). Analysing different demand types (e.g. for home care or equipment) showed the same general upward trend. This was evidence for one of the system conditions described below, i.e. the managers' belief that services needed to be rationed. As a result, the system was moving even further away from being able to deliver to the customer's nominal value.

For the 'Right First Time' measure, it was found that only 52 percent of the cases analysed had achieved the customer's purpose first time round. There were a significant number of examples of service users 're-presenting' (returning to the system) as a consequence of an inadequate response by the service to meeting their needs at an earlier stage. Examples included where users had been sign-posted to services that were never actually accessed by the service user, or where a 'best fit' solution had been provided despite not really being what the service user wanted (for example, a user may have been given adaptive equipment which it would later be found that the user never felt confident enough to use. The adaptive equipment would therefore not have effectively addressed the issue). As a result, the demand from a user was not satisfied and the user would 're-present' into the system at a later date. The numbers re-presenting were evidence that the customer's nominal value had not been met the first time; something in the system had prevented the workers from being able to meet the 'what' and the 'how' of the service user's needs.

Work flow

The team mapped the flow of work in the system at a high level. This showed the work the service had undertaken in order to meet the customer demand. The team identified 18 core flows of work and looked at them from end to end: from the initial customer demand through to the delivery of services to meet these demands. Importantly, this also highlighted points where customers were 'screened out' for various reasons (e.g. not having met the FACS qualifying criteria) and did not receive any service. The average number of value steps across the 18 flows was only 5 percent of the overall work, with the remaining 95 percent being waste from the end user's perspective.

System conditions and management thinking

In Seddon's model, the 'system conditions' are the things that explain why the system behaves in the way it does. Identifying these system conditions can then expose the thinking which is at the core of the problem, preventing the ASC system as a whole from being better able to deliver to the customer's nominal value.

In this system, the main system conditions identified were:

- The need to meet targets and to comply with performance management requirements: the team found that most of the performance indicators in ASC not only drove staff behaviour and additional waste work into the system, but also in many cases led to the system not achieving purpose for the customer.
- The perception that resources are scarce, and that services should be rationed (e.g. the FACS criteria): this was found to cause service users to re-present in a range of cases with more critical needs at a later date, as their original needs (their nominal value) had not been dealt with 'right first time'
- The use of IT was driving the way work was performed: front line staff were found to be spending a significant amount of time in front of their computers recording data into various IT packages which drove the flow of work. The IT had been developed around collecting information for performance management rather than to support operational staff in delivering services.

Each of these system conditions, in their own way, constrained the work of the front-line worker and prevented them from meeting the customer's nominal value. Instead, all were evidence that the managers (guided by policy and procedures) were setting the 'tolerance levels' for the system, rather than allowing the customer to set the nominal value. On this point of perceived scarcity, one interviewee said: 'There's an expectation on public services that people will go for what they will realistically think they can acquire, rather than actually how can you address the needs they've got. We found lots of examples of that in the system where people ring up and give you one need, and when you go and speak to them you often find there were several needs there that haven't been met in the in the past, so they've stopped asking for them. Their expectations [of what the system can do for them] have dropped dramatically'.

This quote suggests that there were hidden demands that were only exposed when the worker was able to ignore the supposed scarcity-avoiding, tolerance level-setting policies and instead to spend time with the service user.

The use of IT was one such indication of the bureaucratic nature of the work being undertaken by social workers, preventing them from spending time with service users in order to better understand their needs. This is borne out by White et al's (2009) work in children's social care (a parallel system to this one) which showed that many social workers are stuck in their offices spending 60–80 percent of their time in front of computer screens, typing up reports to meet targets (*The Guardian*, 19/11/08, White et al 2009). A corollary of this bureaucracy is that front-line workers are not devoting their attention to doing things in the ways that matter to service users. This is seen through the range of forms that have to be completed that have no relevance to the delivery of services: for example in one flow, the service user's ethnic background was asked for 15 times as part of the process. This waste of over specification is an example of the 'loss to society' that Taguchi talks about (Byrne in Ryan, 1988 p. 11). The opportunity cost of not doing things right for the service user can take the form of either of the two elements of a customer's nominal value: 1) not paying attention to what matters to the service user and 2) preventing their demands from being properly identified and met.

Redesign as built around allowing the customer to set the nominal value

As the purpose of ASC had been agreed as 'help people to live the way they want to', the system needed to be designed to be preventative (as opposed to the current system which waited for user's deterioration before intervening), insofar as was possible from this premise. This is aligned with Taguchi's loss function concept. It followed that the new design should respond to all customer demand, rather than turning people away because they were seen to have an insufficient need as had happened before.

The Check team discussed what a perfect ASC system would look like from a customer's perspective and decided:

- Enabling the customer to speak to the right expert as quickly as possible is key to starting the process quickly. This helps eliminate duplication and waste and reduces failure demand from customers who are chasing where things are
- Having a consistent contact through the end-to-end process i.e. from request for help right through to getting the help they need

- The organisation can only really establish a full picture of a customer's needs and what matters to them through face-to-face contact and observation in their personal environment based on an open conversation with them. This includes trying to establish future needs through anticipating any predictable changes in circumstances that might occur so the right support is provided and ideally prevention is put in place to help avoid further deterioration
- In order to meet many customers' needs, it is necessary to use expertise and knowledge from a range of people. The lead contact therefore needs to be able to 'pull' on any support they required as and when it was needed e.g. welfare benefits support
- Frontline staff would be empowered and free to make professional decisions on how best to meet customer needs.

The principles decided upon for this redesign were chosen after discussion with all of the participants in 'Check', based on their knowledge of what they had seen for themselves in the system, and aimed at achieving the above description of perfect from the customer's point of view and ultimately to gain an understanding the service user's nominal value:

1. Build relationships with customers by listening to and clarifying what they want.
2. Anticipate the user's needs (are there things that the social worker's professional expertise would suggest will be necessary for the user in the future).
3. Have access to the right person (right expertise) at the earliest opportunity and the same person throughout.
4. Treat people as valued individuals.
5. Record and measure (proportionately) only relevant information linked to purpose.
6. Support and trust staff.
7. 'Pull' expertise (meaning when you need expertise you don't have, ask and it will come to you and the customer: the case should not be passed on).
8. Continuously improve, don't be afraid to get it wrong.
9. Deliver the right service at the right time.
10. Be honest.
11. Keep things confidential.

These worker-derived principles to guide staff in their interactions with service users were in contrast to the conventional management-imposed policies and specifications. The application of these principles in the work meant that the focus on the whole person and their carers was integral to the new approach, with the result that the service became more person-centred. By freeing up the social workers'/occupational therapists' time, the workers were able to build a relationship based upon trust and understanding with the user. The service user and the expert together were encouraged to identify and provide innovative and creative outcomes to meet the user's need. In effect, these principles were a way of the workers operationalising Seddon's concept of the customer setting the nominal value. In the words of one interviewee: 'Before we were bound by putting the cheapest option in... That is the nice part about it [the systems thinking redesign]. Not being bound by the FACS criteria is good. You can make your own professional judgement, basically'.

The focus of the workers was on promoting choice and independence, developing solutions that often did not cost any more for the service. Everyone who approached the service was dealt with through the application of the above principles of working, which ensured that each individual could feel that they were being treated fairly and equitably. *The whole foundation of a systems thinking approach is to understand individual needs in order to do what the service user wants, hence different approaches to what superficially may appear to be the same problem (from a manager's perspective).* The principles also encouraged workers to work with other health professionals in order to meet the individual needs of the service users and their families.

In order to experiment with new ways of working, the requirement to spend time on reporting performance was removed for the social workers, occupational therapists and other professionals to allow them to focus on these principles. In place of the performance data, the teams were able to establish measures within the work aimed at helping them and their organisation to continually improve, through identifying and removing the 'blockages' that hindered the ability of the professionals to achieve what mattered to the service users. These measures had been decided upon during the Check process as leading the organisation towards the service user's nominal value, i.e. end-to-end time and right-first-time.

Design of first contact in order to understand the customer's nominal value

Studying demand in the Check process had provided knowledge about the expertise that would be required to assess the high frequency, predictable demand. Staff were enabled to develop this expertise such that assessments could be made at the earliest point in the process via a face-to-face meeting. It was judged that this was the best way to gain a proper understanding of the customer's nominal value and to build the necessary relationship with the user. The new design was supported by only one assessment, provision, and review form. This replaced the many forms in the old design, and the reviews of the services provided were determined according to what was appropriate for the user. The information collected was limited to only that which was needed for the provision of the service (which remained subject to refinement with care workers). This meant that the frontline worker was able to spend more time concentrating on understanding what mattered for the service user, identifying their nominal value. Bringing expert staff to the front of the process is often a key feature in a systems thinking redesign.

At the first point of contact, details were taken to establish if the person was already known to the organisation. If so, it was passed to the case worker and if not, the case was passed to the duty worker who would visit to make decisions about need and provision, 'pulling' in expertise from others as required; the case would not be passed over to anyone else. Provision (helping people to solve their problems) in some cases involved sign-posting and/or making creative use of partner organisations such as other local authority departments or voluntary agency/community resources. As one worker said, 'we've been able to take the case on at an early stage, and, working the case with the customer and their carers we've been able to pull in help where we've needed it'.

This practical, in-the-work design meant that there was a focus on the needs of the person as a whole. The approach required the experts (i.e. social workers) to be placed at the front face of the service. The use of one assessment, provision and review form meant that this was an efficient and flexible approach to the work, designed around the nominal value of the service user.

The results of the new system for the individual: Care

The redesigned service was designed so as to be able to do what the service users wanted (meet the customer's nominal value). The purpose of the design was to provide the right support for people in order to maintain their independence, quality of life, and quality of community relations. Experience of the new design had been enthusiastically welcomed by users, with many offering unsolicited praise for the service.

Waiting lists for services had been virtually eliminated: the access team had previously had 150 cases on a 6–8 week waiting list to be allocated prior to the intervention, whilst in the redesign the expert was now closer to the service user and can pull any other expertise needed, meaning that they were now able to deal with cases faster.

Early evidence from the redesign sites suggested that low-cost, early provision was preventing later higher-cost provision, although this would have to be tracked over a longer period of time to be fully verified. Also, pulling support when it was required led to a fuller picture of the customer's needs and financial situation from the outset. All this helped to inform how best to respond to the nominal value of the customer. For example, being able to pull in assistance with user's welfare benefits (a different department) at the beginning of the process has meant that the user's benefits could be maximised. In some cases this meant that no additional funding was required from ASC to provide what was needed for the user.

Increased capability

The new design resulted in the development of different measures which helped the people who did the work to continually improve and identify blockages in the system which were preventing them from delivering against the customer's nominal value. *The new customer driven measures were used to help the people doing the work learn, understand, and improve within the service in their locality and not used as a means to compare and benchmark.* The key measure for the system was decided to be right-first-time. If a high right-first-time measure was to be achieved then the right thing would have been being done for the user and the work would be preventing people from having to 're-present', creating more demand on the system in the future. In the redesign experiment, where all internal and external factors that had been hindering the work had been suspended, the results collected from the redesign showed 90 percent right-first-time. At the same time, the end-to-end time had fallen to an average of 36 days (down from 282 days).

An analysis of Table 3.2 shows the number of the individual service users who were supported that were either known or new to the system, and how many demands they placed on ASC department in each year. The figure shows that the number of times service users are referred has reduced as a consequence of the new approach. In the redesign in this authority, the average number of referrals had shifted from 1.45 presentations of demand in 2007 to 2008 to 1.22 in 2008 to 2009.

Table 3.2 The number of times a service user's demand would reappear in the system

Referrals	07/08 Service users			% Off Total	08/09 Service users			% Off Total	Grand total
	Known	New	Total		Known	New	Total		
Review only	390	0	390	14.9	513	0	513	17.3	903
1 Demand	566	1,363	1,929	65.0	867	1,145	2,012	67.7	3,941
2 Demands	198	300	498	14.2	222	144	366	12.3	864
3 Demands	78	96	174	4.0	51	16	67	2.3	241
4 Demands	34	25	59	1.2	7	5	12	0.4	71
5 Demands	13	13	26	0.5	2	1	3	0.1	29
6 Demands	6	7	13	0.2					13
7 Demands	2	1	3	0.0					3
Totals	*1,287*	*1,805*	*3,092*		*1,662*	*1,311*	2,973		6,065*
% of Total	*41.6*	*58.4*			*55.9*	*44.1*			

* 1,095 Service Users appear in both years, which meant 4,970 individual service users were supported across the two years when they were combined.

The frontline staff attributed these results to the new focus on taking time to build a relationship with clients and, by focusing on getting it right-first-time, this approach was preventing more service users from re-presenting.

These results show evidence of the benefits to performance that can be achieved by concentrating a system on being able to deliver against the customer's nominal value right first time.

Comparative costs of the new design

The research hypothesis was that delivering to the customer's nominal value right-first-time would show a decrease in the organisation's costs, as one element of the reduction in the 'loss to society' as encompassed in Taguchi's Quality Loss Function.

The managers tried to compare the costs of provision in the old and new systems. In order to do this, they used a basic form of Activity-Based

Costing (ABC) whereby notional costs were given for each activity undertaken on the service user's behalf, e.g. for a contact assessment, or for a multidisciplinary meeting. Ten cases were then chosen at random from the 'old' and 'new' systems. Each group of ten included a range of services (home care, domiciliary care, adaptations, and equipment). Calculations of the costs were then generated through an estimation of the average time required for such an activity, then multiplying this by the mid-point of the salaries of the workers involved.

Administration activity costs from the 'old' system (of ten cases taken at random) were calculated from these figures:

- The end–to-end time average was 282 days
- The staff time per case was on average 39.1 hours
- The staff costs averaged out at £923.00 per case
- Average mileage costs of £75.00
- Overall, the gross administration costs were, on average, £998.00 per case

This is a limited data set but it indicates that the removal of FACS criteria does not mean that demand would automatically exceed supply. As mentioned above, many demands that were previously turned away using the FACS criteria came back to ASC at a later date. When this happened, the user was often in crisis and would cost both the care service and the health service much more than if the care professionals had taken the opportunity to spend the right time with the service user at an early stage, thus preventing future demands. This conforms to the expectations of Seddon's adaptation of Taguchi's Quality Loss Function to services.

The average cost of the actual care services provided after generating the administration costs in the above ten cases was £105 per case. At its starkest, this suggested that it was costing approximately £1,000 to provide a service that had an actual cost of £100. This may be a chance reflection of the cases, for often the costs of provision are high and thus would increase any measure of the average. Nevertheless, it is possible to make a comparison of the administration costs between the old and new designs. Following the same method, the costs of administration of ten randomly selected cases in the new design were £134. This was a significant fall in costs, where the savings could now instead be converted back into a greater capacity within the system for providing ASC services.

In some cases, learning more about what the service user wanted actually saved the organisation money that they would otherwise have spent on adaptations. One interviewee described a case where an elderly lady had been originally prescribed a handrail to be put into her house. Ignoring the requirement to get the cheapest of three quotes from contractors, the worker invited one contractor he knew would be able to provide the service along to a meeting at the service user's house. By spending time with the contractor and the occupational therapist, the lady realised that all she actually wanted was a smaller grab-rail that would allow her up a particular step. As the interviewee said, 'Through looking at this carefully and building up a relationship with her, and understanding what she really wants, we were actually spending less money! Previously, the adaptation would have been done by filling in some bits of paper, and we'd have sent out a contractor to put in a handrail.'[2]

The evaluation of the costs of provision in the new design would require a more detailed study of the data over time to make any firm conclusions. However, these early findings suggest that the assumptions in the new design that early provision will prevent or delay later more costly provision are correct. As a corollary, users are expected to maintain their independence for longer. It is also recognised that helping people to live the way they want to is often best achieved through relationships in the community or other sources of provision and would not always incur a cost for the local authority.

Other benefits of the redesigned system

Providing ASC right-first-time has the effect of preventing people in need 're-presenting' as another more complex demand in the future; it also prevents other types of failure demand (such as people calling into the system to say 'where is my Meals on Wheels?' or 'I'm just checking that…') . In some sites where they have had opportunity to experiment fully with a different design in this authority, the failure demand was measured at below 10 percent, down from 74 percent as had been found in Check.

The redesign identifies the value work (assess and provide) and builds roles that do that. Achieving assessments in days and provision in weeks (rather than months or years), released resources, which increased the capacity of the system to do more.

The new design emphasised for managers what competences were required in the new roles and thus was more efficient in terms of training and development. It was anticipated by the managers that working in this less stressful and more motivating environment would lead to less staff turnover and hence less training costs. Reflecting this, one interviewee commented on how much more motivated they were, compared to working in the old system, 'You get a buzz out of being able to say to somebody "you need a piece of kit, let's deliver it to you", knowing fine well that traditionally they would have never got that piece of kit'.

As ASC is a complex, human system, waste was being created through not dealing with user needs properly when they first presented. As a result, needs became more complex and demands on the system were being amplified (one original demand became many unresolved demands). This suggested that the greatest opportunity for intervention would be at the point of first contact: the redesign confirmed this. This also has an impact on the experience of the system by the user, allowing them to perhaps slow or halt the rate of decline in many users' health conditions. This is perhaps an example of the 'unknown and unknowable' consequences of bad management actions that Deming alluded to (1982 p. 121).

These all show the additional 'soft' benefits that could be gained from designing a system which is better able to deal with the customer's nominal value.

Levels of demand

The study of demand in this case study has meant it was possible to measure the number of people receiving services. In fact, demand was found to be stable at 25,000 people (only showing common-cause

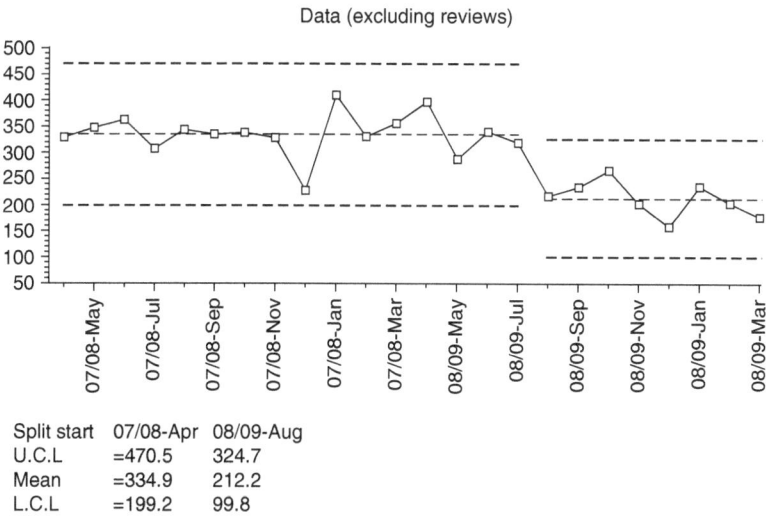

Split start	07/08-Apr	08/09-Aug
U.C.L	=470.5	324.7
Mean	=334.9	212.2
L.C.L	=199.2	99.8

Figure 3.6 Statistical process chart showing demand over time

variation). The chart below shows the number of referrals per month made to the ASC department over 2007–08 and 2008–09. In fact, in August 2008 when the changes were made to the new way of working, a perceptible drop was observed in overall demand (see Figure 3.6).

An associated concern was that removing FACS would lead to a spike in demand. However, the experience gained from experimenting in this authority is that this did not happen. New 'low' and 'moderate' demands were seen to amount to only 7 percent of all demand. Solving peoples' problems in these instances was not costly.

Conclusion

It is noteworthy that, by focussing on the customer's nominal value and effectiveness (better service and an improved service user experience) systems thinking was able to deliver substantial efficiency improvements as a second order result. This is the same continuous improvement experience as witnessed in the Toyota Production System (Ohno 1988, Womack, Jones and Roos 2007). By eradicating the requirement for 'feeding the performance machine at all costs' from this system, there was less of what Taguchi would call the 'loss to society' from not having delivered to the customer's nominal value. Such elements of the 'loss to society' included the costs to the authority (and the taxpayer), the harm done to the independence and well-being of the service user, the concern of the user's carers and family, and the damage to the morale of the staff involved. By focussing on effectiveness (doing the right thing for the customer) and focussing on perfecting the way that the service worked from the customer's perspective, this organisation was able to show greater efficiency.

It was the contention of the staff in this local authority that the use of systems thinking allowed for the government's personalisation principles and policy objectives to be met without the need for the rationing and screening out of services as was previously happening as a result of the FACS criteria. Some of the results included:

- A fall in end-to-end waiting times from an average of 282 days to only 36 days
- A reduction in administrative costs of approximately 85 percent per case from on average £998 to £134 per case (from a sample of 10 similar cases, comparing figures from before and after the redesign of the work)

- 90 percent of cases being dealt with right-first-time, compared to a previous figure of 52 percent

It was also found that government policy of rationing access to care was actually amplifying demand on the service: by making it harder to qualify for care, people's health would deteriorate and the person would 're-present' to the system, meaning that their conditions would often be more costly to treat. After an initial analysis of the data, the numbers of people 're-presenting' in this way was suggested to represent 80 percent of demands on the system. This discovery flies in the face of the cost-control justification for the rationing that was previously taking place in the system.

Through experimenting with the design and management of their work, the authority had been able to remove the requirements to (in the words of the Check team) 'feed the performance machine at all costs' which dominated the way that they used to approach their work. As part of their studies, the authority had discovered that it was the requirement to comply with the specifications promulgated by regulatory bodies and inspection regimes which were the causes of many of their costs and which were behind subsequent poor service to the end service user. As a result, the authority concluded that these specifications (such as the FACS eligibility criteria) should be discarded and further experimentation be allowed in order to demonstrate what results can be achieved in their absence, instead using measures set only from the users' perspective.

Table 3.3 provides a summary of the benefits realised in this case study. These results show that systems thinking can be profitably applied to an ASC department. The benefits of discovering the nominal value of the customer and then designing a system flexible

Table 3.3 Summary of the benefits of the systems thinking intervention

Adult Social Care	Old System	Redesigned System	Improvement
End to end time	282	36	87 percent
Failure demand	74 percent	10 percent	–
Administrative cost	£998	£134	87 percent
Right First Time	52 percent	90 percent	–

enough to deliver against it right-first-time were shown in the results achieved. In fact, in this system, as many of the vulnerable service users were unable to articulate their real needs, the skills of the professional and their carers were required to ascertain exactly what mattered and the precise nature of the demand from the customer. The FACS criteria and the other management requirements for rationing of resources were the main specifications that were preventing the frontline worker from being responsive to the varied needs of the users. Taguchi's Quality Loss Function and the concept of nominal value have therefore been demonstrated to have great relevance to this ASC service.

Notes

1. It is worthwhile mentioning that, in manufacturing, there are parallels to be drawn to the process of finding out the 'voice of the customer' (Hauser and Clausing, 1988).Where this is seen as desirable (e.g. the aspects of design required in a new car door), finding out what customers want is seen as part of the product design process.
2. Additionally, the interviewee was able to discern that the service user was very concerned about showing to the outside world that she had a disability, having been burgled soon after her husband's death. If a handrail had been installed as the original demand required, then it was clear the lady would have felt a great deal of mental discomfort and worry from this new, conspicuous adaptation having been installed into her house unnecessarily. This mental discomfort would have been another example of 'loss to society' caused by not addressing the actual nominal value of the customer.

References

Ashby, W. R. (1956) *An Introduction to Cybernetics*. Methuen: London.

Ashby, W. R. (1958) 'Requisite Variety and its Implications for the Control of Complex Systems'. *Cybernetica*, 1(2), 83–99.

Cabinet Office (2005) 'Transformational Government – Enabled by Technology', The Stationery Office: Norwich.

Chapman, J. (2002) *System Failure*. Demos: London.

Checkland, P. (1981) *Systems Thinking Systems Practice*. Wiley and Sons: Chichester.

Commission for Social Care Inspection (2008) 'Cutting the Cake Fairly: CSCI Review of Eligibility Criteria for Social Care', CSCL: London.

Davis, F. W. and Manrodt, K. B. (1996) *Customer Responsive Management*. Wiley Blackwell: London.

Deming, W. E. (1982) *Out of the Crisis*. MIT Press: Massachusetts.

Deming, W. E. (1994) *The New Economics: For Industry, Government, Education*, MIT Press: Massachusetts.

Department of Health (1998) 'Modernising Social Services', The Stationery Office: Norwich.

Department of Health (2002) 'Local Authority Circular 13: Fair Access to Care Services', The Stationery Office: Norwich.

Hauser, J. R. and Clausing, D. (1988) 'The House of Quality'. *Harvard Business Review*, May–June, 63–73.

HM Govt (2008) 'The Case for Change – Why England Needs an New Care and Support System' Dept of Health Publications.

Jackson, M. (2000) *Systems Approaches to Management*. Kluwer/Plenum: New York.

Jackson, M. (2003) *Systems Thinking: Creative Holism for Managers*. Wiley and Sons: Chichester.

Jackson, M., Johnstone, N. and Seddon, J. (2007) 'Evaluating Systems Thinking in Housing'. *Journal of the Operational Research Society*, 59, 186–197.

Johnson, H. T. and Kaplan (1987) *Relevance Lost: The Rise and Fall of Management Accountin.*' Harvard Business School Press: Massachusetts.

Levitt, T. (1972) *Production-Line Approach to Service*. Harvard Business Review, September–October, 41–52.

Lochner R. and Matar J. (1990) *Designing for Quality: An Introduction to the Best of Taguchi and Western Methods of Statistical Experimental Design*. Quality Resources: New York.

McQuade, D. (2008) 'Leading Lean Action to Transform Housing Services'. *Public Money and Management*, 28(1), 57–60.

Neave, H. (1990) *The Deming Dimension*. SPC Press: Tennessee.

Office of the Deputy Prime Minister 'A Systematic Approach to Service Improvement Evaluating Systems Thinking in Housing' Sept 2005 (in particular Jackson's Appendix on Systems Thinking)

Ohno, T. (1988) *Toyota Production System*. Productivity Press: Portland, Oregon. Translated from Japanese original, first published 1978.

Porter, M. (1985) *Competitive Advantage*. The Free Press. New York.

Ryan, N. (ed.) (1988) *Taguchi Methods and QFD: Hows and Whys for Management*. ASI Press Dearborn: Michigan.

Saunders, M., Thornhill, A. and Lewis, P. (2007) 'Research Methods for Business Students' Prentice Hall 4th Edition. London.

Seddon, J. (2005) *Freedom from Command and Control*. Vanguard Press: Buckingham.

Seddon, J. and Caulkin, S. (2007) 'Systems Thinking, Lean Production and Action Learning' *Action Learning Research and Practice*, 4 (1), April 2007, special issue: 'Lean Thinking and Action Learning.'

Seddon, J. (2008) *Systems Thinking and the Public Sector*. Triarchy: Axminster.

Spear and Bowen (1999) 'Decoding the DNA of the Toyota Production System'. *Harvard Business Review*, Sept–Oct, 97–106.

Vanguard Education (2001)'The Vanguard Guide to Using Measures to Improve Performance' Vanguard Education: Buckingham

Vargo and Lusch (2004) 'Evolving to a New Dominant Logic for Marketing'. *Journal of Marketing,* 68 (January 2004), 1–17.

White, S., Wastell, D., Broadhurst, K., Peckover, S., Davey, D., and Pithouse, A (2009) 'Children's Services and the Iron Cage of Performance Management: Exit the Street Level Bureaucrat, Enter the Good Soldier Svejk?'. *International Journal of Social Work*.

White, S. (2008) 'Drop the Deadline' *The Guardian*, 19 November.

Womack and Jones (1996) *Lean Thinking: Banish Waste and Create Wealth in Your Organisation*. Simon and Schuster: New York.

Womack, J. P., Jones, D. T. and Roos, D, (2007) *The Machine that Changed the World*. Macmillian: New York. First published 1990.

Yin, R. K. (2009) *Case Study Research: Design and Methods*. Sage: London 4th Edition.

4
Seeing Shared Work within a System

Rhian Hamer and Sarah Lethbridge

Following the Gershon Report (2003), a public sector Shared Service Centre (SSC) was created to bring about joined up services and service transformation. This paper evaluates the Change Programme that was established in order to facilitate this transition in terms of success and sustainability. It also examines, to some extent, the services that the organisation provided. It discovered limited results from the original change programme, evidence of 'failure demand' (Seddon, 2005) and a dislocation from customer experience of the services offered. The paper then goes onto to describe a new programme of activity which has been designed to reshape the organisational system in order to increase the amount of 'right first time' work delivered to customers.

Introduction

In 2003, Sir Peter Gershon was commissioned to conduct a review into public sector efficiency. His report *Releasing Resources for the Frontline: Independent Review of Public Sector Efficiency* made proposals to deliver sustainable efficiencies within both central government and the wider public sector. His argument was thus:

> The public sector typically spends two or three times more per employee than the private sector on human resources which can increase to as many as six times more even within central government. Procedures need to be simplified, standardised, shrunk and shared, with central Government put into 'clusters' of departments that would share services.

The report presented a series of proposals which, if adopted, were forecast to deliver over £20 billion of efficiencies in public spending by

2007–08. At the heart of his proposals, and key to the facilitation and support required for departments to deliver their efficiencies were change agents – teams of specialists who have experience in delivering business transformation and identifying cross-departmental opportunities.

In December 2006, Varney published a report that strengthened the findings and recommendations made by Gershon. Customer's expectations of government were changing; citizens expect faster services and are constantly making comparisons between public and private sector service offerings. Varney focussed his review on integrating frontline service delivery in order to save government, citizens and business time and money.

He looked at opportunities for improving channel management within the public sector – the way in which services are delivered (e.g. contact centre, face-to-face, e-channels). He argued that providing more joined up services would reduce duplication and would bring about increased service satisfaction across the sector and hence result in efficiency savings. The report highlights the need for service transformation, benefits of which include greater personalisation, improved speed of service, improved convenience for end users, and generally a more consistent quality service delivery within the public sector.

A key message runs through both the Gershon and Varney reports: Business transformation is fundamental to the creation of a culture focused on delivering efficiency savings, and driving improvements.

Following the Gershon report, a public sector Shared Service Centre (SSC) was created to bring about joined up services and service transformation. This chapter evaluates the Change Programme that was established in order to facilitate this transition in terms of success and sustainability. In order to evaluate the success of the change programme, it will be necessary to examine the particular problems experienced within a shared service environment and determine whether appropriate steps have been taken to overcome these issues. It concludes by proposing a new way of thinking about work as a solution to the problems experienced. Finally, it discusses the early results of this shift in thinking.

Case study organisation

The organisation in question is a National Shared Service Centre which is part of Her Majesty's Prison Service (HMPS) and is located in Newport, South Wales. Her Majesty's Prison Service serves the public by keeping in custody those committed by the courts. Its duty is to look after inmates with humanity and help them lead law-abiding and useful lives in custody and after release. The service employs 55,000 staff, is responsible for circa 80,000 inmates and costs £2.5 billion per year to run. There are 128 public prisons located throughout England and Wales, the larger jails employing as many as 600 officers and smaller ones as few as 50.

HMPS 'went live' with Shared Services for Finance and Procurement at the end of April 2006, followed by Human Resource Services in October of that year. In line with its growth strategy, from February 2008 the Centre started to provide the same services for the Home Office, making it the first example of two separate government departments receiving their service from a single provider. The centre is now a multi-service line and multi-customer Shared Services, employing over 1200 staff over two sites.

Shared Services operates four main business areas: HR, Training Services, Procurement, and Finance, all of which are supported by a Contact Centre and IT function. Table 4.1 gives an indication of some of these service offerings.

The change programme

HMPS Shared Services began implementing Six Sigma as soon as the organisation was created in order to facilitate change and create an

Table 4.1 Shared service centre business streams and service offerings

Business stream	Service offerings
Finance	Expense management Accounts receivable Banking services General ledger reporting and accounting Cash management
Human resources	Recruitment and vetting Resource management Pay and contracts
Procurement	Contract management Invoice processing Requisition to procurement Accounts payable
Training services	Continuous professional development of staff Training admin and support Training delivery

environment of improvement. Strong emphasis was placed upon the need for standardised process maps, control plans and a robust measurement system to enable the organisation to measure performance against a set of pre-defined service level agreements (SLAs).

As part of the induction process every employee attended a one day Six Sigma awareness training day, designed and delivered by the internal change team known as the Process Improvement and Quality department (PI&Q).

Selected staff were then given the opportunity to progress to yellow or green belt status. Selection was based upon set criteria regarding analytical skills, ability to use certain software packages such as Microsoft Excel and the identification of a business 'problem' which forms part of the delegate's training project. Support was provided to the delegates throughout their training project via mentoring and coaching by assigned PI&Q team members.

During 2006–2008, Shared Services trained 26 staff to yellow belt in two waves and 40 employees underwent green belt training, in 3 waves. Yellow belt training was designed and delivered by PI&Q, Green Belt training was delivered by an external training supplier. Focus of the six sigma projects was functionally specific, with dedicated members of the PI&Q team aligned to and physically located in a particular business area. Typical projects titles included:

- Reduce the number of duplicate utility bills from x to y by end of Jan 2008
- Reduce the number of non-'transactable' IT forms from x percent to y percent within 3 months
- Reduce the average time spent by agents in after-call work within the HR contact centre by 10 percent by October 2008

Project identification took three forms: bottom up projects suggested by staff working within the business streams; top down from middle/senior management and those proposed by PI&Q as part of business reviews and rapid improvement events. Project selection and prioritisation was the role of the senior management team via steering committees which were held on a quarterly basis. The purpose of the steering group was to drive the projects and resource in line with the strategic direction of the organisation.

Throughout the lifecycle of the projects regular review sessions were held with the sponsors to ensure the project was on track to deliver pre-specified targets and to overcome any roadblocks. Green and yellow belts were mentored by more senior members of PI&Q, providing advice and guidance where needed.

Upon project closure, financial assessment was conducted and project savings were noted. During the first two years of its operation, Shared Services implemented over 250 six sigma projects with projected *potential* savings of around £500K.

At the end of December 2008, although project targets and savings were being met, the Shared Service Centre's Change Department considered the effectiveness of their change initiative limited and exclusive. In line with 'Plan, Do, Check, Act' the Change Team approached the newly appointed Director of the Shared Service to gain approval for commissioning a piece of work to establish whether the perception of the Change Department was a reality and an important question had to be answered: how successful has the initiative been?

Learning from the literature to inform research

It has been widely acknowledged that the success of change programmes is sporadic. In 1992, an Ernst and Young report stated that three quarters of transformations fail. In 1993, Hammer and Champy suggested that at least 50 percent of business improvement programmes are deemed to be long term failures and up to 70 percent fail to achieve all of their intended benefits. Nearly 20 years later and unfortunately the statistics tell the same story; a recent study, published by the Center for Creative Leadership, reported that between 66 and 75 percent of all public and private change initiatives fail (Kee and Newcomer, 2008) a figure corroborated by Beer and Nohria (2001).

Much research has been conducted to determine the key components of sustainable change. The results of a survey involving 93 private organisations, listed the most frequent implementation problems; inadequate training, insufficient resources, lengthy implementation times, and poor leadership were amongst the top ten (Lucey et al. 2005). Radnor et al. (2006) support some of these findings in their analysis of lean programmes within the public sector. They identify some of the key issues as 'lack of resources to implement changes, resistance to change from staff and management, post rapid improvement event lack of ownership for the improvement activity, lack of management and staff commitment throughout the change process and slow natural pace of change in the public sector' (p. 3).

The value of recognising the effect these elements have on change programmes lies in developing a series of enablers designed to counteract them. In *Staying Lean*, Hines et al. (2008) posit that 'level 5' leadership qualities, excellently developed and deployed strategy and a positive, engaged workforce are all essential within a transformation programme if the organisation wishes change to be sustainable. A fully communicated, successfully deployed strategy is required to focus change activity in line with business needs and the overall vision of the organisation. In order to test this, Hines et al. (2008) suggest that the organisation should ask itself the following two questions to gauge organisational knowledge and engagement with strategy:

1. Can all the people in your organisation clearly articulate what your strategy is?
2. Can they demonstrate what they are doing in their normal job to help the organisation achieve this strategy?

Crucially, for improvement activities to offer the most benefit to the organisation and consequently, be more likely to be sustained, they too must be directed in line with the strategic objectives.

Excellent leadership is required to communicate strategy and to develop people. Good leaders have a 'guiding vision, passion and integrity. They are innovative, people – focussed and are willing to challenge the status quo'. (Hines et al. 2008 p. 9). People are seen as the critical element in lean transformations (Emiliani, 2007). An 'engaged workforce' should be 'engaged' in continuous improvement as a key part of the role. Continuous Improvement is the responsibility of everyone in the organisation (Crosby, 1995). These concepts are visually illustrated in Figure 4.1, the 'Sustainable Lean' iceberg model (Hines et al. 2008).

The 'Iceberg model' posited by the authors could be seen as a pictorial depiction of the innovative, learning organisation Senge (1990), Spear and Bowen (1999) and Liker and Meier (2007) regard as the reason for Toyota's success. Toyota provides workers with the right environment (trust, respect, and a positive approach to problem solving) and the tools (Socratic reasoning, A3s) in order to make continuous improvement happen.

The Shared Service environment itself offers new challenges. Work occurs at a distance from where it is initiated and required. Seddon argues vehemently against the 'Shared Service' model pursued by Government stating that it 'it is doing the wrong thing wronger' (Seddon, 2006). He argues that the Government have been seduced into creating huge factories of work in pursuit of economies of scale, but that shared services

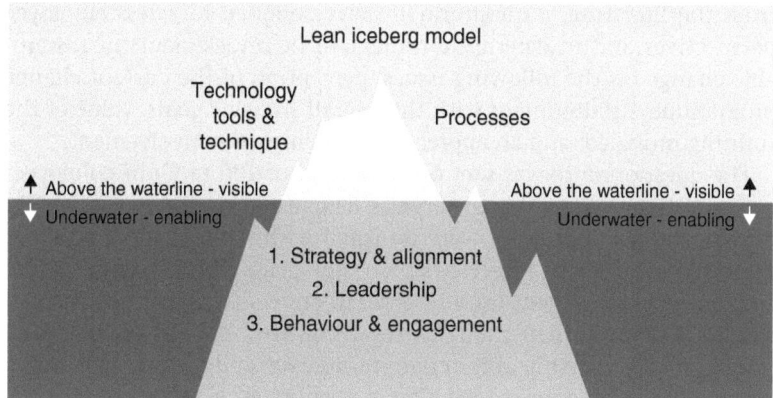

Figure 4.1 The sustainable lean iceberg model
Source: Hines, P, Found, P. Griffiths, G and Harrison, R., 2008.

are not solving a customer's problem right first time, but are actually becoming more and more entrenched in 'failure demand.'

Seddon introduced the concept of 'failure demand' as a volume of work which is strangling public services. Failure demand is the failure to do something right, or to do something right for the customer. The existence of failure demand overburdens public services, increases the amount of processing time, and creates serious dissatisfaction for customers. In order to evaluate the success of the change programme at the Shared Service Centre, it is essential that these concerns are explored in greater depth to determine whether such concerns are founded. To what extent does 'failure demand' comprise the work of the SSC?

Evaluation of the SSC Change Programme needs to ascertain the extent to which the factors which we have just discussed (leadership, an engaged workforce, excellently developed and deployed strategy, and an awareness of systems thinking) exist within the organisation. This evaluation becomes the 'Check' part of the 'Plan Do Check Act' cycle. It will be essential for the senior management team at the SSC to 'Act' on the result of its findings, directly targeting the shortcomings of the original change programme within the next iteration of improvement work.

Methodology

In order to evaluate the change programme, qualitative and quantitative research methods were employed. Following knowledge gleaned from the literature, a questionnaire was designed to assess employee perspectives and to determine the extent of key elements of sustainable change on the following issues: perception of the current change programme, its alignment with the overall strategic goals, value of the training provided, and an appreciation of employee involvement.

The questionnaire was sent out to a total of 100 random employees across all areas of the Shared Services. The response rate was very high, 82 percent of the surveys were returned within the 2 week window. Eight semi structured interviews then took place with key senior members of the management team in order to understand their perspectives and to determine their thoughts regarding strategic alignment of the programme, ownership of the change initiative, and the perceived value of the change programme. Finally, in keeping with an action research approach, the experiences of the change team within the SSC itself will form an important aspect of reflection.

Results

Collating results from the employee questionnaire offered an opportunity to gauge widespread opinions about the change management programme at the SSC. Figure 4.2 shows us that very few respondents felt that involvement in improvement initiatives was not beneficial with most respondents (61 percent) indicating that they thought that they were 'very beneficial'.

However, when questioned about the role that improvement should play within day to day work, responses indicated that it was more a 'nice to have' than something that everybody should get involved with (Figure 4.3) suggesting that continuous improvement was not seen as an essential part of working life.

Indeed, the success of the six sigma programme itself was debatable. Figure 4.4 shows the profile of the green belt delegates following their training. At the end of 2008, 57 percent of the trained delegates had completed their first project (a requirement for certification), 35 percent then went on to complete a second project and a staggering 80 percent were deemed inactive – i.e. not involved in any improvement activity whatsoever.

Over 60 percent of those taking part in the questionnaire had not been involved in any of the six sigma projects to date. When coupled with the fact that, at the time of the survey, over 250 projects had been completed throughout the organisation, it can be said that improvement teams were either quite small or made up of the same members.

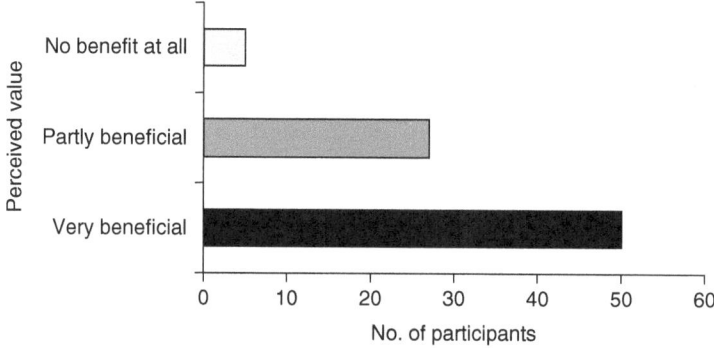

Figure 4.2　Perceived value of change initiatives to the business

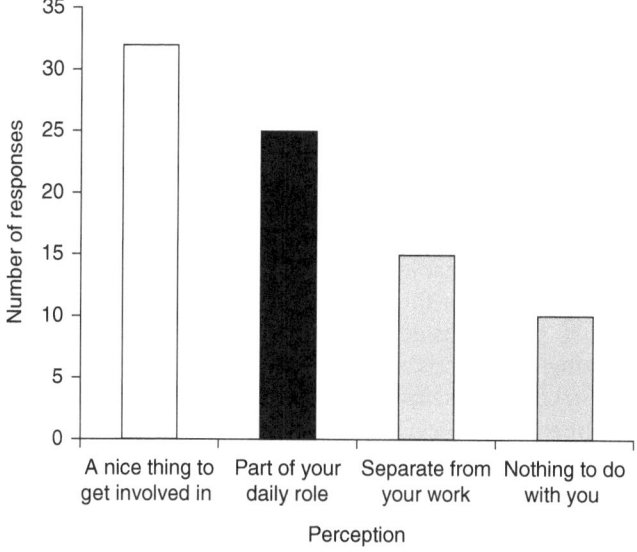

Figure 4.3　Respondents' perception of process improvement

Sustainability of the improvements was also questionable. Very often projects deliverables were not achieved and the metrics around which the improvement targets were set began to slip as the processes returned to 'old' ways of working or attention shifted to a different part of the

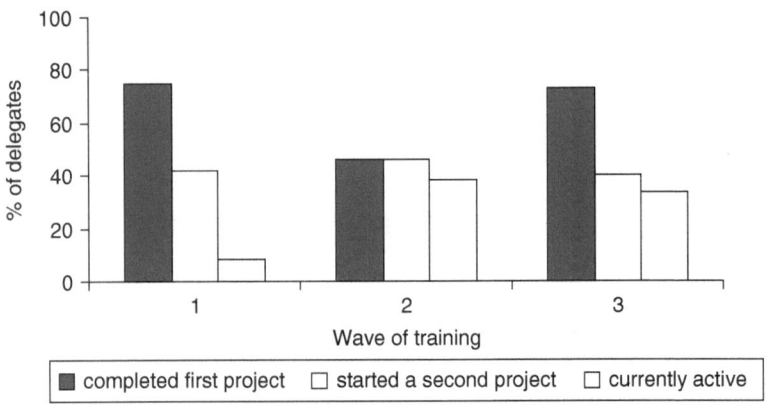

Figure 4.4   Green belt training success

organisation. Furthermore, PI&Q felt as if they were single-handedly driving improvements which were not aligned to business needs and therefore limited in their effectiveness.

Results from the questionnaires and interviews indicated that very few employees were able to articulate the strategic objectives and explain how the work they carry out links to the overall business goals. Question 14 asked: 'What are the SSC's current strategic objectives?' A staggering 70 percent of the respondents failed to comment at all, the remaining 30 percent listed between 2 or 3 points which made reference to 'being a world class service provider', 'growth' or 'cost reduction'. No two comments made were the same and none of the responses clearly outlined the strategic objectives.

Semi structured interviews

The interviews offered an opportunity to examine the 'position' of continuous improvement within the senior management team. Members were asked to comment on their understanding of current SSC strategic objectives, their involvement in the development of these objectives and how well they perceived their services were aligned to this strategic direction.

Responses were mixed. It was clear that the senior management exhibited inconsistent views towards the high level strategic direction of the organisation. Organisational aims expressed ranged from 'unit cost reduction for continuous improvement especially through the use of technology' to 'to be in the upper quartile in terms of performance' to 'comply with legal and HMPS policy requirements' and 'to grow to satisfy customers'. If the senior management team were unclear about the core strategic aim of the organisation, how could continuous improvement be deployed to deliver against these goals?

In terms of 'owning' continuous improvement, 6 out of 8 of the senior management team felt that they were responsible for improvements to their service streams, but they were yet to work cross functionally across value streams to examine improvements within the 'system' or to empower staff to a significant enough extent to be able to enact improvement ideas on their own (Figure 4.5).

Seven out of the 8 interviewed perceived that of the improvement activities that had taken place to date, there had been some success, however, the responses to the open question 'what value have you seen from the change programme' told a different story. One response commented that it 'looks good on the walls, whether or not that translates as change on the ground, I don't know. I suspect that it's not driving change' and another stated that 'the value that was billed to be achieved was not delivered. The software solution led to an increase in headcount which was not expected'.

The interviews collectively suggested that systems thinking (Seddon, 2003) was not prevalent in the SSC. Senior managers were very focussed on meeting their own service stream SLA's and there was little evidence

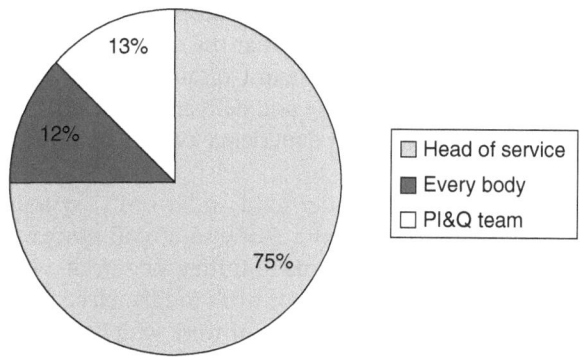

Figure 4.5   Responsibility for process improvement at SSC

of cross functional improvement initiatives. Many of the responses were contradictory. Although the PI&Q team were valued, it was clear that there was some misunderstanding around what the team were there to deliver and how improvement should be communicated to all. Statements such as 'people see this [continuous improvement] as a chore on the shopfloor' and that the six sigma training 'did not engage the staff, nobody has said that they are glad to be a green belt' support this.

Observations from PI&Q Team

So, what was the experience of the team which had been established to bring about these improvements? At the point of evaluation, it was clear to team members that a change in direction was needed in order to reenergise the change programme and ensure that their work had clear commitment from Senior Management.

When leading improvement work, it became clear that there were some problems about the way employees approached work within the SSC. For example, employees had little visibility of what their customers were actually experiencing in terms of end to end service performance. They needed to understand the true nature of customer experience of SSC transactions. For example, in March 2008 analysis showed that 10 percent of calls answered within the HR contact centre were employees questioning receipt of overtime claims and associated monies into their bank account. The internal, published SLA for completing such claims was 5 working days from receipt of a clean form, which often led to employees phoning 6 days after they handed their claim to their line manager for authorisation to 'chase' receipt of the form and to query when the payment was to be received. Analysis of the work from a 'systems thinker' point of view proved that the end-to-end cycle time was on average 42 days; this took account of authorisation, internal and external mailing systems, sorting and delivery, processing and the pay roll process itself. The customer experience proved very different to the 5 days internal SLA.

It was crucial that the SSC understood work within a system. The work that they transact is one part of a true end-to-end process. For example, an employee in a prison needs to initiate the work, seek authority and then communicate the need for work to take place before the SSC even discovers the work. Many pieces of work which 'arrive' at the SSC have already been in existence within the 'system' for some time. This actual, end-to-end time was not measured.

One measure that was collected was that of 'transactability' – a term created to describe the amount of work which is 'rejected' because it does not reach the SSC in a form that can be processed. Seddon would see this work as a type of 'failure demand'. The customer has approached the organisation to 'solve their problem' but because the work did not appear in the format that the SSC were able to process, it was forced out of the system, only to reoccur as rework. Employees needed to understand that the root cause of the 'transactability' problem needs to be explored and then consequently eliminated in order to improve the service experience to customers. Failure demand was prevalent in the SSC for example, 95 percent of calls into AP and Finance Contact Centre were from suppliers chasing non payment of an invoice.

Summary of evaluation

In light of knowledge gleaned from the literature regarding successful organisations that deliver value in public services and organisations that instigate and manage sustainable change programmes, a number of conclusions can be drawn about the condition of the SSC at the time of evaluation:

From a systems thinking point of view:

- Failure demand was prevalent e.g. suppliers chasing payment of invoices
- Service teams did not appreciate the true end-to-end experience for customers
- Primary focus on meeting service level agreements, measures did not account for complete transaction time
- There was little regard for establishing 'root cause' of systemic issues

Evidence of strategic thinking within the organisation:

- Employees had little visibility or knowledge of the organisation's strategy
- Improvement initiatives were not linked to the organisation's strategic aims
- Measures were not customer focussed but were bound within the four walls of the SSC

Leadership

- Continuous improvement was not championed by leaders and remained the sole responsibility of PI & Q team
- Middle management sought strong leadership from the top with regards to reflecting continuous improvement behaviours

Employee engagement:

- Few 'continuous improvement' role models existed within the organisation
- Detachment from responsibility for true customer experience
- Concern that there was still some way to go to change perception of continuous improvement as a 'bolt on'

The results of all of the primary research conducted dictated that a new approach was required. Shared Services needed a change programme which was both successful and sustainable in its approach, one that looked at work as a system, focussing on solving customers' problems right first time.

The new programme

It was decided that the new programme should target middle managers to encourage ownership of improvement at a level where they were knowledgeable about their work and possessed the authority to make change occur. A pilot course was developed specifically to address the following aims:

- Help Service managers and team leaders to appreciate the true experience of customers using their service, not just adherence to SLAs
- Understand the amount of 'failure demand' that they experience, establish the root cause of these failures and work to eliminate them
- Appreciate that continuous improvement should not be a 'bolt on' 'nice to have' activity, but should be the responsibility of everybody
- Realise where their services are not delivering value to customers and work to resolve issues experienced.

It was particularly important to specifically address the senior management team's thinking on this journey. They needed to realise that internally focused SLAs were not accurately reflecting customer experience and could actually drive the wrong behaviours. Consequently, PI&Q has spent much time working with the SMT in order to facilitate this change in thinking.

The core element of the programme focuses on Service Managers appreciating the need to make processes simpler for customers. Pointless bureaucracy must be reduced in order to solve customers' problems completely. Managers were tasked to understand the amount of rework that existed within the services that they manage and seek to reduce it.

The cross functional composition of the programme facilitated the sharing of improvement ideas and encouraged thinking about the organisation as a system. Colleagues were able to discuss how work moved across different departments and work together to make work more efficient and effective.

Critically, the key thrust of the programme was to clearly communicate that continuous improvement needs to be part of their everyday working life. Senior management have reinforced this message in the development of a new set of values including 'continuous improvement is the way we do things around here'.

Conclusions

Evaluating the change programme at the SSC clearly demonstrated that there was a need to reconstruct and reinvigorate continuous improvement within the organisation – a continuous improvement approach which supported staff to concentrate on solving customers problems the first time they approached the SSC. Dislocation from the customer and measurement systems focussed on optimising parts of the end-to-end process comprised a system condition which obstructed workers from being able to deliver the best services.

The most important aspect of this realisation is that the SSC's senior management team took steps to support a new way of working. The new change programme in the SSC's main mission is to clearly communicate the power of continuous improvement and to help staff to be able to answer the two critical questions posed by Hines et al. (2008) – a clear understanding of what the organisation's purpose is and the ability to affect their role in order to deliver that purpose.

Providing the necessary thinking and methods in order to answer those questions is the crux of the new Service Management programme. Five cohorts of service managers have now experienced the programme and although there is still much to be done, there is a definite acknowledgement that it is no longer acceptable not to see work within a system.

Focussing solely on 'transactability' and meeting SLAs (that are not customer focussed) leads to customer dissatisfaction and a failure to deliver services against purpose. The organisation's measurement system is currently being redefined in order to focus on what matters to customers. The senior management team is working very hard to develop a clear, succinct strategy, and the right system conditions to support a new way of working within the SSC.

Challenges remain regarding the 'detachment' of work from point of origin, but as employees understand this challenge more, they can do more to address it. The important aspect of the new programme is that a problem solving ethos and continuous improvement culture is being supported at every level of the organisation. Staff are experiencing the benefits of incremental change and are positive about their ability to keep improving the service that they offer their customers.

References

Bateman, N. (2001). *Sustainability: A Guide to Process Improvement*. Lean Enterprise Research Centre, Cardiff University: Cardiff.

Beer, M. and Nohria, N. (2001). *Cracking the Code of Change*. Harvard Business Review. Boston: Harvard Business School Press.

Crosby, P. (1995). *Quality Without Tears: The Art of Hassle-Free Management*. McGraw-Hill.

Emiliani, B. (2007). *Real Lean. Understanding the Lean Management System*. The Center for Lean Business Management (The CLBM) Kensongton, Connecticut.

Ernst & Young, (1992) *International Quality Study: Best Practices Report*.. Ernst and Young and American Quality Foundation: Cleveland.

Gershon, P. (2004) *Releasing Resources for the Frontline: Independent Review of Public Sector Efficiency* HM Treasury.

Hammer, M. and Champy, J. (2003) *Reengineering the Corporation, A Manifesto For Business Revolution*. HarperBusiness. London.

Hines, P, Found, P. Griffiths, G and Harrison, R. 2008. *Staying Lean. Thriving, Not Just Surviving*. Lean Enterprise Research Centre.

Kee, J.E. and Newcomer, K.E. (2008) (Fall) Why do Change Efforts Fail? What can Leaders do About it? The Public Manager. Accessed 2 July 2009. http://findarticles.com/p/articles/mi_m0HTO/is_3_37/ai_n30939464/

Liker, J.K. and Meier, D. P. (2007). Toyota *Talent. Developing Your People The Toyota Way*. McGraw Hill.

Lucey, J. Bateman, N. & Hines, P. (2005) 'Why Major Lean Transformations have not been Sustained'. *Management Services: Journal of the Institute of Management Services*, 49 (2), 9–13.

Radnor, Z., Walley, P., Stephens, A. and Bucci, G. (2006) *Evaluation of the Lean Approach to Business Management and Its Use in the Public Sector*. Scottish Executive Social Research.

Seddon, J. (2005) *Freedom from Command and Control*. Vanguard Education Ltd:. Buckingham.

Seddon, J. (2006) 'Speaking against the Motion, Shared Services will Deliver the Greatest Efficiency'. Institute of Revenues, Ratings and Valuation Annual Conference, October 11th 2006.

Senge, P. (1990) *The Fifth Discipline: The Art and Practice of the Learning Organization*. Currency Doubleday: New York.

Spear, S. and Bowen, H. K. (1999). Decoding the DNA of the Toyota Production System. *Harvard Business Review*, (September–October), 97–106.

Varney, D. (2006) *Service Transformation: A Better Service For Citizens and Businesses, a Better Deal for Taxpayers*. HM Treasury.

5

Improving Performance Throughout a Housing Supply Chain: Portsmouth City Council's Systems Thinking Transformation

Brendan O'Donovan and Keivan Zokaei

Introduction

This is the story of Portsmouth City Council's housing management service, which has transformed the way it works by designing services against local demand. The results achieved are outstanding, making the targets the service was set by central government look unambitious in comparison. These results were achieved through the adoption of a different way of managing and delivering services based on systems principles, in spite of the approach's inherent antithesis to the 'command and control' ideology of many of the previous government's public sector reforms. As a consequence, there is a large disparity between the results achieved by the service and the way it has been treated by the inspection bodies and the government's Department for Communities and Local Government (CLG). Inspectors have not been able to see the usual hallmarks of their 'best practice' guidance at work in Portsmouth and thus have scored the services poorly against their inspection criteria.

What is particularly noteworthy about this housing management service is the way that many organisations are collaborating to deliver within one system, as systems thinking has been adopted all along the supply chain for housing services. Private sector suppliers have managed to halve the costs per repair in this way, working in true partnership with the council to deliver excellent services to local residents.

Together, they have designed a service that can undertake repairs on the day and the time that the customers want them. Through working to stocking their repair vans against what was predictably required in a certain area, the contractor firms now spend less than 25 percent of what they were spending on stock before. On top of this, there has been real innovation within the system, in both designing IT in-house to support their work (developed at a fraction of the cost of conventional off-the-shelf housing IT packages) and in starting a new logistics arm which supplies materials to tradesmen exactly when they are required.

Background

Portsmouth City Council is a unitary local authority on the English south coast. Unlike many local authorities, Portsmouth City Council's housing management service has retained direct ownership and management of its housing stock. It is the largest social landlord in the south Hampshire sub-region with over 17,000 tenanted and leasehold dwellings, representing 18 percent of tenures in the city. Portsmouth's council homes play a significant part in meeting regional housing priorities. The department has an operational budget of £80 million and comprises a staff of approximately 600.

What makes Portsmouth's housing service different is that it has become one of the public sector's most fully-developed examples of systems thinking, with systems principles in use across the whole department and integrated into the service's supply chain. After studying their service in detail, the Portsmouth housing managers made an informed decision to set their own measures rather than follow the centrally set targets and specifications which were imposed on their service. The results achieved by these services have been outstanding. The outcomes achieved as a result of this deviation from national guidelines have been remarkable and have attracted some favourable coverage for the service from the press, academics, politicians, and think tanks, but ironically the housing service has also drawn criticism from the Audit Commission and the government's Department for Communities and Local Government (CLG).

Before the change to using systems thinking, the department was scoring highly against conventional government measures of success. In July 2006, the Audit Commission rated the council as doing very well: the Housing Service had been rated as three out of a possible four on the council's Comprehensive Performance Assessment (CPA) scorecard and had achieved Beacon status. The repairs and maintenance service was rated as a 'good 2 star service, with promising prospects for improvement'. Performance indicators from 2005 to 2006 had shown voids (empty residences) being turned around and re-let within 30 days, which placed them in the upper quartile of performance for other local authorities.

Despite this strong record of achievement in the service, local councillors would regularly accost the Head of Housing Management, to tell them that their surgeries were filled with residents complaining about having to wait for up to a year for repairs to be attended to in their council houses. This ran contrary to the council's survey results which said that 98 percent of residents were happy with the service they were providing. This rating was based on the answers to questions such as whether workmen were friendly (and smiled) when they came round, and whether they had cleaned up after their work. The council's KPIs (Key Performance Indicators) showed that repairs were being carried out within budget and within the time targets as specified centrally and measured by the Audit Commission. The Head of Housing Management decided to investigate further what was causing these contradictory messages. Discussions with colleagues led to attendance at a systems thinking course, and the subsequent decision to apply systems thinking to the council's housing repairs service.

As a result of extensive demand analysis throughout the department, Portsmouth has found that managing in accord with centralised housing policies was resulting in poorer services for local residents which were rife with inefficiencies. The council's experience shows that a 'one-size fits all' approach to centrally specifying housing services does not draw strengths from local individuality and identity.

Initial investigations into these services from the service user's point of view made managers discover things which shocked them. In essence, they found that the services were designed remotely against theory rather than experience, they were heavily target-driven and as a result they were disconnected from what end-users actually wanted. The result of such a design was that people in the organisation 'left their brains at the door' when they arrived at work as they were required to blindly follow processes rather than be able to react to a customer's direct needs. As a consequence, people were afraid to be wrong or simply to highlight issues that arose and instead people cheated the system so as not to show failure to comply. Workers were left being motivated extrinsically instead of wanting to do a good job for intrinsic reasons.

Not only were the department's services to users actually very poor, but the people working in them had their creativity and originality suppressed to such a level that the work had little chance of improving. There was a general overriding attitude that people's main concern at work was to 'keep the inspectors happy'.

It was discovered that throughout the organisation there was a culture of 'learned helplessness' and safety in numbers: people were penalised for taking responsibility or pride in their work in the way it had been organised. Instead, they were dependent on a daily menu of targets against which they would need to perform, conforming to the expectations of their managers who continued to serve up these targets. They expected processes and policies to come from the corporate centre or from the government. Work primarily consisted of satisfying the managers higher up the chain rather than focusing on the delivery to the housing tenants. If there were problems identified, it required a project or a 'Best Value' review before they got fixed.

Thankfully, by following systems thinking and the respective intervention methods, the organisation has been able to turn this performance on its head. For example, the council's housing repairs and maintenance service is now delivering far beyond previous expectations (with repair costs having been halved on average, and with the service achieving much higher levels of user satisfaction) as it is fully designed around meeting customer needs instead of primarily achieving targets. People are now encouraged to be imaginative, to alert managers to issues, and work together to solve them. As a result, the need for extrinsic motivation has been replaced with the intrinsic motivation of doing a good job for the customer.

Processes selected for improvement activity

As a department, Housing is able to operate with a degree of independence from other council functions, as it is governed by a different set of legislation. This level of independence has been a contributing factor to the success of the systems thinking interventions in Portsmouth's housing department: there has been more freedom for the department to take its own line with regards to dealing with the 'system conditions' which have been discovered to prevent the department from delivering excellent service. The managers within the department have embraced the autonomy their status affords them and taken on the responsibility of improving all services for residents.

Within the housing department, there have been interventions in the following services:

- Repairs and maintenance
- Planned maintenance
- Applications, voids (empty properties), and lettings
- Cleaning and grounds maintenance
- Housing allocations
- Rental income
- Sheltered housing services

As the housing department's systems thinking team became more established, it has also been given the opportunity to become involved in redesigning the council's corporate human resources and recruitment systems.

The first intervention: Reactive housing repairs

Portsmouth first applied systems thinking to the repairs and maintenance service. As this has been the system which has now been running for the longest period of time, systems thinking has been integrated throughout the service and down into the layers of suppliers and contractors in the system. Whilst housing repairs systems thinking interventions have been documented elsewhere (e.g. ODPM 2005, McQuade 2008), this intervention is especially noteworthy for the level of integration throughout the whole supply chain.

The learning acquired during the council's own intervention led the contractors who provided the housing operatives to also acknowledge some of the problems within their own organisations. The council subsequently required all of its housing contractors to become systems thinking organisations in their own right – a challenge to which four of the five original contractors responded. The fifth did not change its systems and as a consequence did not have its contract renewed. (See below for accounts of individual contractors in this system.)

In the first intervention in the repairs and maintenance service, the initial process began with selecting a team to undertake 'Check', the first stage in the 'Check-Plan-Do' approach (Figure 5.1). These were representatives from throughout the repairs and maintenance service structure. All had different skills, perspectives, and personal attributes.

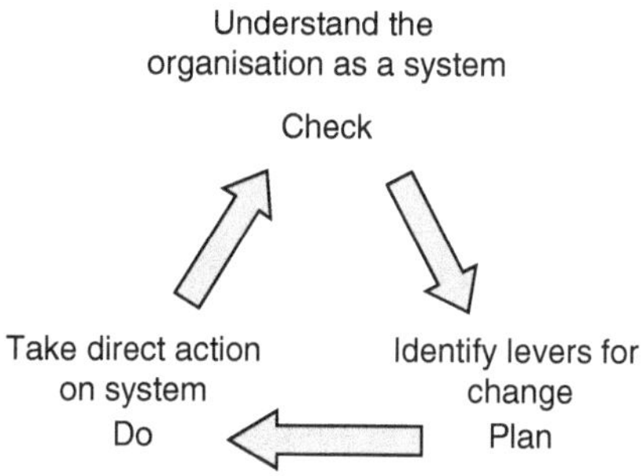

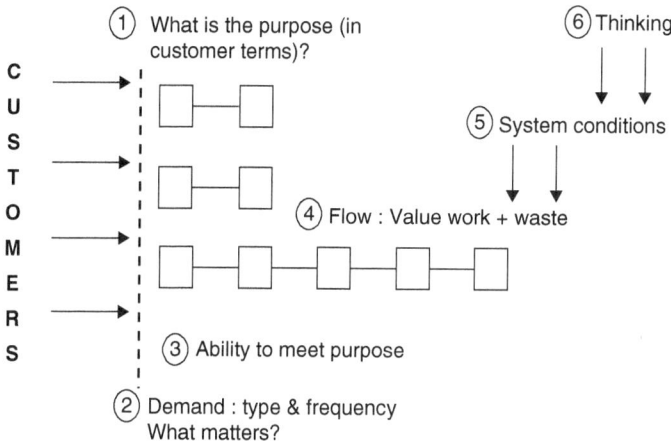

Figure 5.1 Systems thinking 'Check-Plan-Do' approach
Source: Seddon, 2005, p. 100.

Figure 5.2 Improvement work/project: the stages of 'Check'
Source: Seddon, 2005, p. 112.

Importantly, all were selected as they were known to be respected by their colleagues, but all joined as willing team members.

The members of the team were made fully aware that this was not just another training course. All representatives were workers selected from within the existing system (i.e. there were no representatives of residents in the Check team). Team leaders and senior managers who had management responsibility for this service were also required to contribute to the Check process (see Figure 5.2) and to assist when necessary.

Purpose

After an initial period listening in on actual customer phone demands that were coming into the call centre, the Check team were given an insight into the customer's housing requirements from the system. It was then that the team agreed what they thought was the purpose of the system:

'Do the right repair at the right time'.

This was in contrast to the existing de facto purpose, which paid more attention to the council's strategies and business plans, focusing on measures of activity for elements of work over specified time periods or budget expenditure or against Key Performance Indicators (KPIs). This simply defined purpose was continually validated by the other findings of Check, listening to customer demand and feedback throughout the intervention.

Demand

Step two of Check focuses on demand and what matters to the customer. The team were required to gain a first-hand understanding of demand by listening to calls and spending time observing at the housing department's seven area reception counters. The team found that perceptions about the type and frequency of demand (for example that it was unpredictable) were not borne out empirically. In fact, the opposite was true: demand could be shown to be predictable throughout the times of the day, month, and year. Figure 5.3 records demand into the system and shows that the mean number of jobs per week was quite predictable over time.

When this demand was broken down and analysed, it was discovered that value demands were typically in the form of 'I need something fixed' (i.e. they were first time requests for repairs to be carried out) and presented themselves as 'can you help / advise me?' Failure demands were of the type 'I've reported a repair – not heard anything since?', 'You've been to repair but it's not finished', 'I'm unhappy with the quality of the repair', 'I need a repair fixed again (repeat request)' or 'you said you'd be here but you didn't turn up'. Failure demand was discovered to be running at 60 percent of all demands.

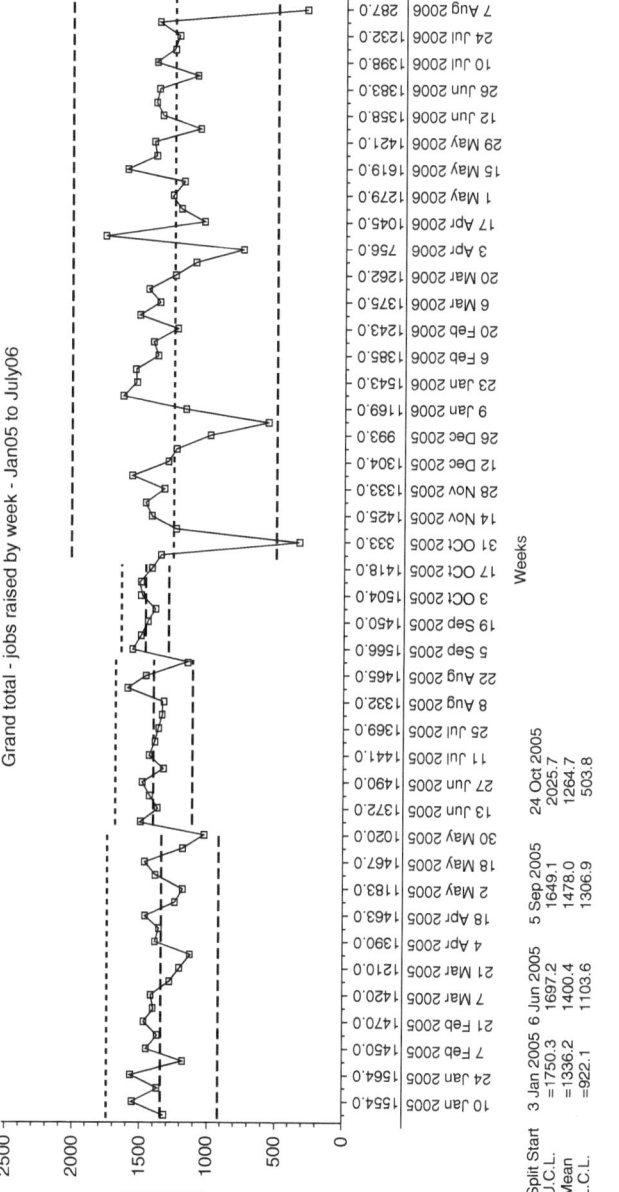

Figure 5.3 Demand coming into the repairs and maintenance system

The second stage of demand analysis involves deciding what matters to the customer, which should inform how the organisation responds to the value demands they were receiving. The elements that the team agreed upon, based on the experience of listening to demand included that repairs would be completed which were:

- completed in one visit;
- would stay fixed, not requiring another call to the council for a further repair;
- undertaken either;
- as soon as possible; or
- at a time convenient for the customer.

As these were the agreed measures of what mattered to the customer, the team investigated the capability of the service to deliver against these measures.

Capability

The statistical process charts (SPCs) as shown in Figure 5.4 demonstrate the capacity of the system to deliver end-to-end repairs. Predictably and reliably, it was taking 98 days for a repair to be completed, with a mean time of 24 days.

When investigating the current system's capability to deliver against the purpose of the system, the Check team were astonished to find that none of the current measures in use related to this purpose. All of the existing measures variously related to activity, budgets, or performance against KPIs as specified centrally and monitored by the Audit Commission. A great deal of the staff's time was spent collecting unreliable, inaccurate data. As soon as the team began to piece together end-to-end times for repairs based on historical records, they found that actual times were substantially longer than the previously reported conventional measures suggested. In fact, they discovered that 15 percent of all repairs required four or more visits for a job to be completed.

Flow

When the flow of work was mapped from the initial receipt of a request for a repair through to completion of the task, the team was able to construct a diagram showing all of the various steps and decision points in the flow (Figure 5.5).

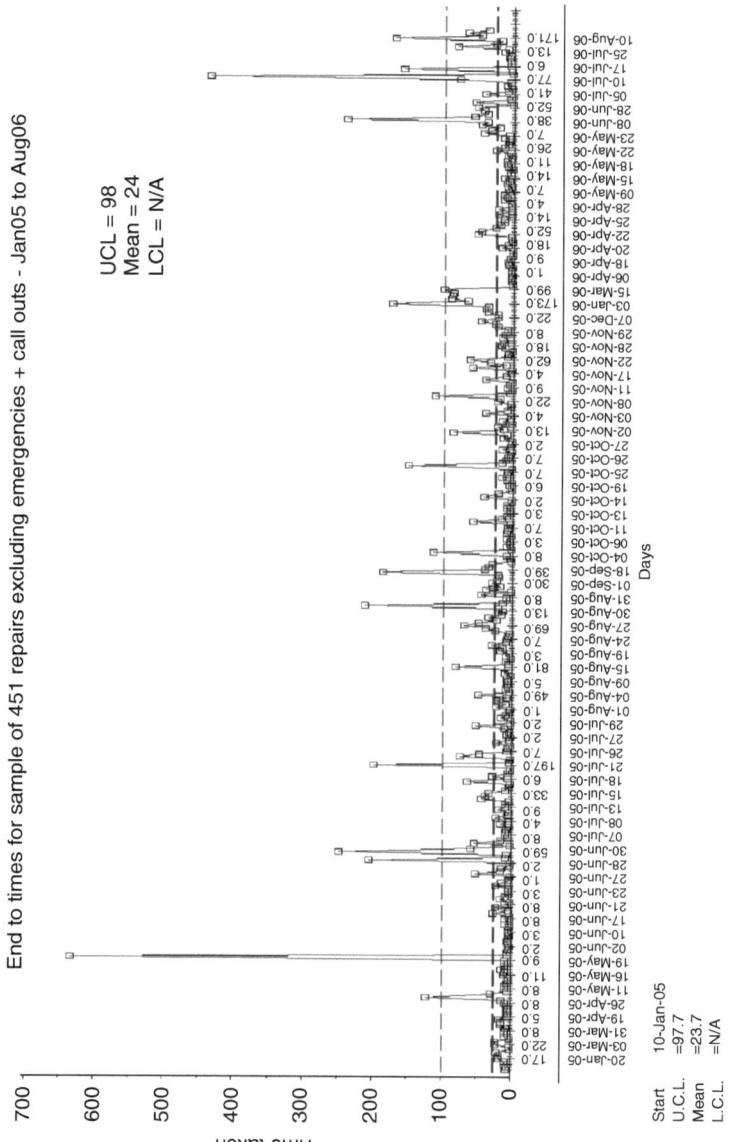

Figure 5.4 End-to-end times taken to make repairs

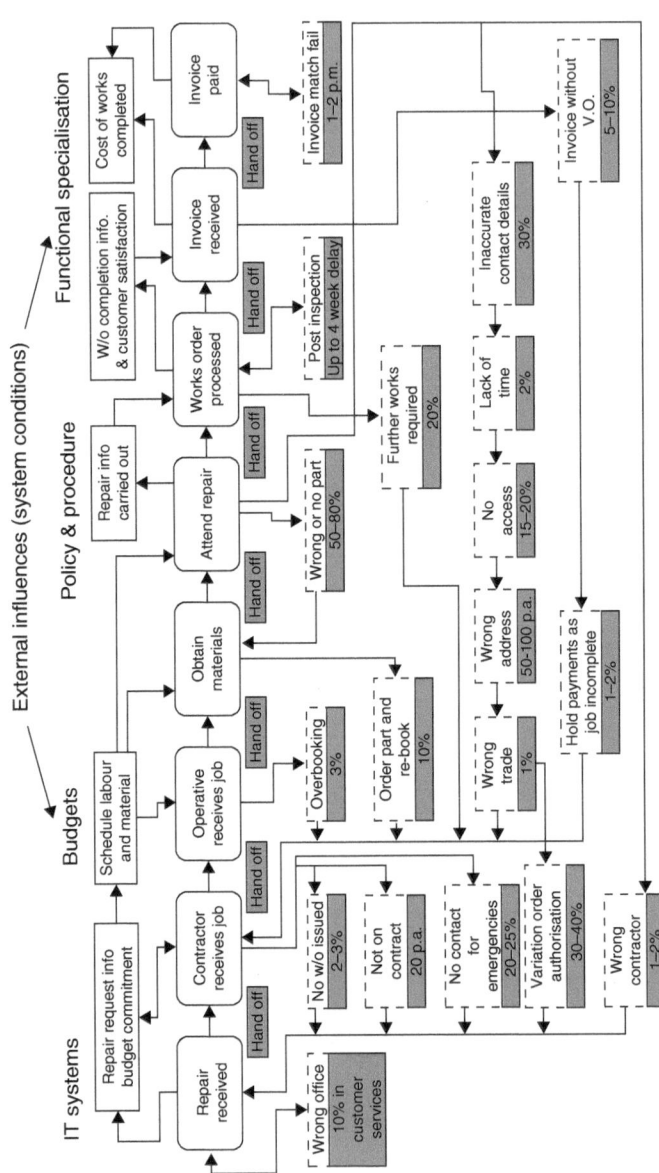

External influences (system conditions)

Functional specialisation

Policy & procedure

Budgets

IT systems

Figure 5.5 System picture of the housing repairs system, including the system conditions – the manifestations of the thinking behind the design and management of the syste

The Head of Housing Management was able to relate a personal anecdote which demonstrated what he had learnt during his time exploring the flow. He had quickly found that, in reactive repairs, things were only fixed when they became really bad. On one occasion, he was following a plumber in the work for a day when they went to see a repair of a leaky tap. The plumber's behaviour was driven by the fact that each patch had an amount to spend each month, so repairs were only undertaken on the problem which was reported, not any other problems which presented themselves.

He quickly realised that this was causing calculations about costs based on false economy. For example, changing a washer in a dripping tap cost approximately £25, whereas it was estimated that changing a tap would cost £35 plus parts. The plumber would therefore be required to do the cheaper piece of work and change the washer to provide a short term fix to a leaky tap. Of course, the plumber usually did this because he was doing the job that was specified on his ticket, instead of being free to use his professional expertise in order to put in the 'right' repair. However, when the check team drilled down into the information, they found that washers had been changed on average 3 times, with others being changed 13 times rather than replacing the tap.

Therefore, a more realistic calculation was to compare the cost of 3 x £25 + £35 + parts = £105 + parts against the one-off cost of £35 + parts. The right repair would have been the right one for the property and the customer. In learning this, there were clear lessons about the capacity for improvement in both performance as judged by the customer and costs accrued over time in the planned and reactive maintenance services.

Figure 5.5 illustrates the decision points in the repairs process and the system conditions which were affecting these decisions. The main steps in the process of making a repair and getting paid using this system are marked in bold. Below this line of steps are records of how often problems occurred which would either require repeat visits or delays.

Going through the Check process meant that the team learned many more detailed things about the system along the way. These included:

- The pre- and post-inspections which were being carried out on repairs by non-technical staff to check that operatives were doing their job in fact were adding no value from the customer's point of view, and were only causing delay and inconvenience to customers.

- The repairs that were not being fixed first time led to increased and unnecessary activity in the system and were a cause of failure demand.
- Staff members throughout the system were focussed on meeting budgetary targets and improving KPI scores. In turn, this was aimed at protecting the organisation's good external image but not their image with their customers.
- The service was completely reactive rather than proactive, and good service was being judged on the quantity of complaints received rather than direct feedback from residents. There were leading questions in the surveys.
- Repairs were not treated as a high priority for non-technical staff in the organisation.
- There were very high levels of waste, which were resulting in poor utilisation of staff resources and skills.
- Contractors were carrying little or no stock on their vans as the contractor companies did not trust their workmen not to use the stock to do other things. As a result, contractors had to make several visits to a residence in order to complete a single job.

Previously, the Head of Housing Management said that the department had spent money on marketing the department to their residents, which in turn was aimed at keeping the results of satisfaction surveys high. Through Check, it became obvious that this was adding no value for the customer, and hence there is now no money spent on marketing. Instead, the only satisfaction data would come directly from tenants who have received a service and was used in order to improve, rather than to demonstrate performance.

System conditions and what was revealed about management thinking

In the systems thinking model, the system conditions are the things that explain why the system behaves in the way it does, and the major conditions identified here (as shown in Figure 5.5) were summarised as being:

- the targets imposed on the system from above in the form of policies and procedures;
- pressure to meet budgetary requirements (managing costs);
- the division of work into functional specialisms; and
- the inflexibility of the IT systems.

A further condition identified by the senior managers in the organisation was the acknowledgement that the success of people's careers depended on gaining awards and improving the council's corporate star ratings. These are commonly treated as indications of good service when in fact they drive behaviour which makes service worse.

Identifying these system conditions exposed the thinking which was underlying the way the system had been designed and managed. Changing this thinking is the ultimate goal of systems thinking: 'Thinking' governs the 'System' which in turn governs 'Performance' (Seddon 2005 p. 10).

The system conditions were discovered to have many distorting effects on the actions of people in the system:

- Managers are driven to pressure call centre workers to change job classifications on the system. This is because emergency jobs are perceived to be more expensive for the organisation than routine repairs, and also in order for the organisation to be able to achieve its quotas of job categorisation targets which it has to report to the Audit Commission.
- One job from a customer could be turned into multiple jobs on the system. Jobs would be cancelled and restarted, for example when a tradesman was not able to gain access to a property. This would avoid missing time targets for jobs.

Every job thus looks as though it has been completed on time and in budget to meet the targets. However, in fact the customer is experiencing poor service, feeding the demand into councillor's surgeries which originally triggered the Head of Housing Management's investigation into systems thinking. Pressure would also come from the targets that try to get housing organisations to do more planned than reactive repairs. When the system is studied, it is found that the arbitrary separation between planned and reactive repairs drives the organisation to refuse to replace kitchens for some people because they are due to have a planned replacement in six months' time, whilst for others they find perfectly good kitchens being ripped out to meet the planned schedule. The team learned that it was actually better to make replacements when it is best for the property and the resident, based on knowledge of when such repairs are necessary.

Many of these system conditions resulted from the need to comply with centrally dictated specifications that emanated either from government or inspection bodies. For example, the Decent Homes

Standard promoted by the government forced the council to undertake a cyclical maintenance programme, whether properties required replacement of kitchens, bathrooms, etc. or not. These requirements to comply were identified as external threats to the sustainability of the service redesign.

Redesign

The redesign incorporates the experimentation with both the new methods of working in the 'Plan' stage and then the roll-in of the rest of the organisation in the 'Do' stage. The team agreed that the crucial 'value' steps in the process, as judged form the customer's perspective, were:

- ensuring access to the property;
- diagnosis of the problem; and
- completion of the repair.

By experimenting with new ways of delivering repairs to residents, several further discoveries were made about ways to improve the process:

- Specific timed repair appointments enable better access.
- Accurate diagnosis of what is required to conduct a repair can only effectively be carried out at the property by the correct operative (for example electricians for electrics).
- For repairs undertaken and appointments, customers need to be seen as individuals, all having different 'nominal values' (different things that matter to them about how they receive a service) which need to be acknowledged. This is in contrast to the way times were previously categorised as emergency, urgent, or non-urgent which did not allow for customers' needs to be differentiated.
- Each visit to a property is an opportunity to maintain it and prevent future repairs. The operative would always ask 'Is there anything else that requires doing?' when on a job.
- Repair decisions must be separated from other influences such as budget or other targets and practices. The priority becomes 'do the right thing for the customer/the property' rather than 'put in the cheapest solution'.
- The correctly skilled operative must be given the right resources to undertake the right repair.

- By working in a responsive manner, the council not only satisfies their customers but is also proactive in looking after its housing stock to prevent future maintenance issues.

The key points are:

- Portsmouth now carries out repairs when it finds them rather than batching them into 'planned maintenance programmes';
- if a kitchen or boiler needs replacing, on the professional judgement of the tradesman, then it is carried out at the convenience of the customer;
- it has been established that the previous 'textbook' component life-times (which were driving the 'planned' maintenance programmes) are not borne out in practice; and
- the external decorations programme was previously carried out in a five year cycle; however, the changes made mean that currently 40 percent of the blocks have external decorations over five years old and yet there has been no increase in related demands.

Many of these changes required new ways of working for suppliers as well as the council's staff. This highlights the importance of working on the whole system and changing the thinking (and contractual obligations) of suppliers to improve overall performance.

The changes of redesign aimed at only doing the value steps (assess, diagnose, complete repair). This required the best expertise to be placed at the front end of the process. Housing staff now focus on questioning customers about their problem to ensure the right skilled tradesperson is sent to the repair at a convenient time for the customer. This has resulted in the right tradesperson making the first assessment of the problem.

As they progressed through redesign the team realised 'wouldn't it be good to fix everything which may need fixing at the same time?' This required a change of thinking: they now had to trust the tradesman to act professionally, which the system was not doing previously. This change of thinking was not easy for all managers to accept. To produce normative learning they had to take some of these managers through the process several times before they were comfortable. They found that they had to overcome people's suspicions that tradesmen would be workshy by seeing the work circumstances for themselves.

Table 5.4 compares the results of the work before and after the intervention.

Costs: Using the old separation of reactive and planned repairs, reactive costs increased as the tradesmen discovered latent demand from houses which had become run-down over time. Portsmouth reported that costs per job had actually fallen by 7 percent through the redesign, although there was now more demand being discovered by the tradesmen as they asked 'is there anything else that needs fixing?' However, by removing the arbitrary distinction between planned and reactive repairs, Portsmouth discovered that the savings from the planned maintenance budget more than funded the increase in reactive repair costs. Overall repair costs have remained virtually the same for the past three years. See Table 5.1.

Overall, the housing repairs service (including both planned and reactive repairs) has managed to reduce the spend on repairs from £35 million in 2007–2008 to a budgeted spend of £32 million in the current year. Hence, overall costs for repairs have fallen over this time. The council has now worked on (and succeeded in) reducing the end-to-end time for all repairs, including those only discovered on arrival at the property.

Systems thinking integrated along the supply chain

Similarly, dramatic results have been achieved by the individual contractor organisations in the supply chain who have adopted systems thinking methods. One contractor was shocked to discover that they previously had no measures which were helping them to learn and improve in their business. Their aim has now become a true end-to-end repairs business. This contractor has substantially increased their capacity whilst only using the same staff, moving from a mean of 85 jobs per day of up to 225 jobs per day, allowing for additional work to be taken on from other council contractors. They also found that the key element of being able to complete a job right-first-time was the requirement for suitable van stock.

Table 5.1 Repairs and maintenance expenditure

Year	Spend (£)	
2007–2008	35,114,720	Actual spend
2008–2009	29,369,358	Actual spend
2009–2010	30,141,880	Actual spend
2010–2011	31,775,000	(Budgeted spend)

Table 5.2 Results of the process before and after improvement

	Before	After
Purpose	Manage all activity in order to meet the targets and keep down costs	To do the right repair at the right time
Measures end-to-end	Predictably 24 day average for all repairs	Down to average of 6.9 days to fix the originally reported repair and 11.2 days to fix all repairs identified at the property
Failure demand	60 percent	Currently running at approximately 14 percent for repairs
Customer satisfaction	Old MORI measures were running at 98 percent for the service. However, true satisfaction rates with the old repairs service, as measured in the process of Redesign, was 6 out of 10	Repairs service has been recorded at an average of 9.9 out of 10 for the past 6 months
Right First Time	Not measured	91.8 percent
Appointments kept	Not measured	77.5 percent

This was designed against demand for parts, meaning that the vans were replenished at the right frequency. They decided that fast access was required for parts that could not be carried as van stock so as to improve their right-first-time measures. The four service providers for Portsmouth City Council all measured the time spent collecting parts and found an average of 1.8 hours per day, per tradesman.

As a result, one of the contractor company's directors decided to set up their own rapid response parts supplier company. Parts are sourced and delivered directly to the site of a repair. This means that the skilled tradesman does not have to leave the site to collect additional parts, causing inconvenience to the resident. Before the intervention, end-to-end times (at this particular contractor) averaged around 13 days with an upper control limit of 59.1 days. Post-intervention, these figures have fallen to an average of 2.8 days, with an upper control limit of 9.4 days. Figure 5.6 shows how, as a result of making a first time fix (reduction of visits per job from 2.9 to 1.9), the average repair cost per job fell from £258 in April 2009 to £114 in October 2009.

Another set of measures from this contractor showed that failure demand was now running at 9.4 percent, when it was at 25.6 percent before, meaning that the council is better able to respond to customer repair requests.

Planned maintenance

An example of systems thinking applied to a non-transactional service can be seen in the council's approach to their 'planned' maintenance service. Conventionally, housing departments are required to undertake 'planned' maintenance (i.e. not 'reactive' to resident demands for repairs) to their estates according to schedules which are based on the expected shelf-life of materials, to comply with external 'best practice' standards/regulations or to comply with a 5-year rolling programme of replacements. Essentially, repairs were being made on a 'push' basis: they were taking place in accordance with a predetermined plan rather than based on true knowledge of the actual need for replacement of items such as boilers, kitchens, or bathrooms. There were qualified building surveyors who were nominally employed to understand the needs of buildings, but who it was discovered were actually spending all of their time ensuring compliance with the maintenance schedules rather than assessing actual need. Their professional skills were being neglected.

The key realisation in this system was to recognise that, for the purposes of the intervention, the house itself should be treated as the 'customer'. The 'Check' team in this intervention (made up of representatives from the council and its contractors) decided upon the purpose of the system as being to 'maintain and improve our property'. As the intervention proceeded, this purpose was validated by the housing residents.

Being a non-transactional service, all of the work was outbound. In order to study the existing system, the team tried to explore what data they already had about the state of their properties. They collected information from:

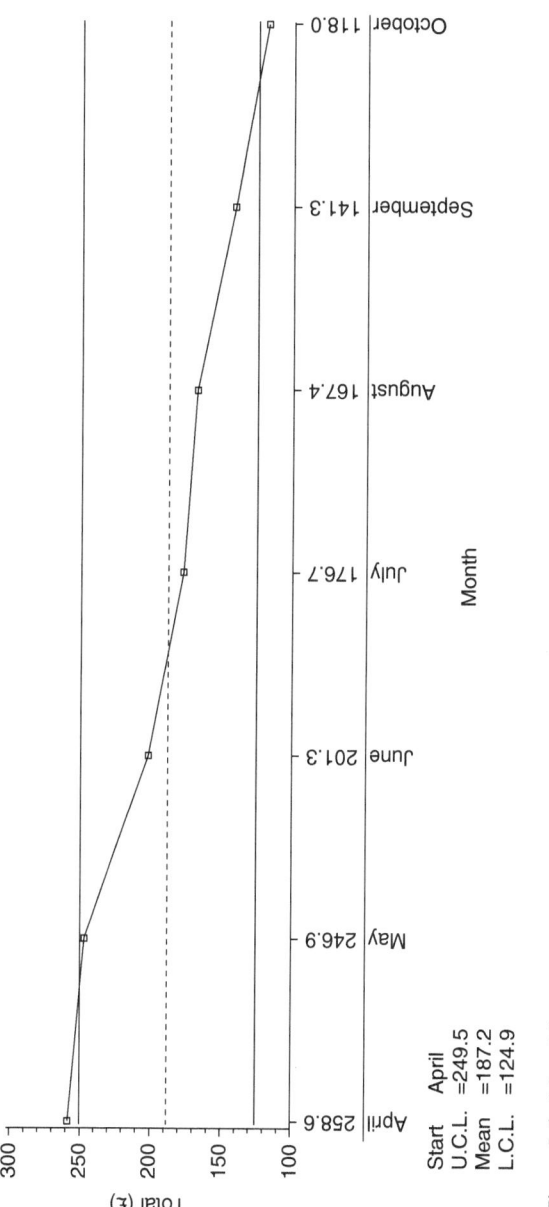

Figure 5.6 Monthly average repair cost 2009–2010

- their 'Stock Conditions Database', which said what work had been done, the expected shelf-life for products and when things were last replaced
- by asking housing officers in different geographical areas what they knew regularly needed replacing in the housing blocks in their area
- by talking to tenants about what predictably went wrong in their houses

The team then decided to focus in on one block of flats and understand the likely shelf-life of the components based on the local circumstances, reengaging the building surveyors' professional skills in order to assess the true state of parts of the property. They looked at where repairs were historically being made. They listened to tenant feedback and studied the predictable demand for reactive repairs (e.g. they discovered predictable problems with boilers in the block). They also looked at the number of recalls that happened after planned maintenance had taken place to look at the actual shelf-life of components, and where there were 'too many miles on the clock' for reactive repairs to be carried out. The building is now treated as the customer in the cases of planned maintenance – the reactive repair demands are an indicator of problems, but the individual tenants do not or cannot tell tradesman when all the boilers in a block need servicing, or when a roof needs replacing rather than being continually patched up. Ascertaining what the building would say if it could speak requires 'listening' to the patterns of reactive repair requests in an area, the use of on-site surveys and identifying any patterns in the costs, types, and frequencies of particular repairs. Such work (e.g. replacing a leaking flat roof with a longer lasting pitched roof) can lengthen the life of the element by another 25 to 30 years but will also reduce the number of reactive repairs.

Now, having determined that the purpose of the system was 'maintaining and improving the properties', the measures for doing the work were decided upon. These were:

- End-to-end time for completing work
- Did the property stay fixed?
- Did the work go as it should have?
- Did the maintenance programme fix the problem (e.g. if an electrical problem was fixed in one flat, did that only lead to electrical problems in other flats in the block)?

This work helped them to redesign their stock conditions database, filling it with information based on the data collected so that they could be more accurate in predicting the expected life of a product such as a new kitchen, from which they were able to better schedule future work.

As in the redesigned reactive repairs system, where the repair operative asks the tenant 'is there anything else that needs fixing whilst I am here?', the operatives were required to judge if there was anything else that the property needed before finishing their job. For example, if there was scaffolding constructed to do painting work, they would examine if the guttering needed to be replaced or whether TV aerials or chimney stacks needed repairs. This iterative learning approach meant that the scope of work quite often was broadened from the initial starting point for the maintenance.

By studying their work in detail, they exposed huge amounts of inefficiency in the old system that they have now eliminated:

- The work being undertaken was being governed by plans, policies, and procedures decided at a distance from the actual work, thus meaning they were inaccurate.
- Many of the replacements based on textbook shelf-life of components were unnecessary – they were ripping out and replacing perfectly functioning kitchens or bathrooms in line with these policies and procedures which was wasting large amounts of money.
- Surveyors now were reengaged with their work, actually using professional judgement and building knowledge about whether components required replacement rather than enforcing work to arbitrary plans as they had previously.
- The old IT system was preventing the reallocation of money in the repairs system meaning inefficient use of millions of pounds.
- Planned maintenance schedules were supposed to start on the first day of the financial year (April) whilst the budgets were not being agreed upon until June. This was leading to a waste of time for the surveyors who were planning and scheduling schemes that did not take place.

This is an example of an outbound, non-transactional service working from purpose, deciding the appropriate measures and then experimenting extensively with the best method for conducting the necessary work. Once this new method was shown to be robust, the other sites and staff were then all successfully 'rolled-in' to this new way of working so that now the whole council and its suppliers work in this way.

Planned maintenance schemes are now undertaken as a way of 'turning off' or reducing recurrent reactive repair demand by addressing the priority maintenance issues. Due to the small number of planned schemes that PCC have needed to run 'end-to-end' through the new system it was difficult to separate planned from reactive maintenance in terms of both measures and costs. Now, each planned maintenance project is treated as individually and on its own merits (following Check-Plan-Do on a small scale), where the team provide a maintenance service to address the needs of the particular estate/block/street/building as required. The old traditional planned maintenance programmes based on targets and limited by budgets were made up of cyclical contracts where windows, heating, bathrooms, or kitchens were replaced according to a schedule and programme based on dates. This led to the replacement of items which were continuing to function well and which in many cases the resident was happy with.

The pooling of reactive and planned maintenance budgets has meant that overall, the housing department has greater flexibility to do the right thing for the customer and the property. For example, a scheme which is planned to take place is to replace 800 windows in an estate and replace them with newer plastic windows. This is in reaction to recurrent customer demands for repairs to faults with the present older windows. Previously, such a scheme would have had to have been broken down and scheduled over time e.g. 16 schemes of 50 windows at a time in order to fit with budgeting requirements. The team also knows how often things such as external redecorations are required to properties based on experience – they now can undertake this work more frequently (every year) than they used to (every 5 years) in some blocks, where others do not need redecorating for longer periods.

Conclusions – the achievement of effective local service delivery

Portsmouth City Council's approach has had a dramatic impact on the provision of housing services. Customers are experiencing the benefits, as is shown by the customer satisfaction data collected after a repair has been made. The service continues to learn by listening to customer demands and ploughing information back into its services. The process has required the support and commitment of staff across the organisation, as well as residents, councillors, and service providers.

In Portsmouth's experience, centrally controlled housing policy dictates the quality of service provision. Regulations and incentives, of either additional funding, targets, or corporate assessments, work to the detriment of the customer. They also hinder individuality, stifle progress, and prevent local housing authorities from allocating resources to those most in need. In Portsmouth's opinion, local housing authorities have the capability and local knowledge to provide the best service to customers and should be empowered to do so.

References

Office of the Deputy Prime Minister (2005) '*A Systematic Approach to Service Improvement Evaluating Systems Thinking in Housing*', The Stationery Office: Norwich.

McQuade, D (2008) 'Leading Lean Action to Transform Housing Services', *Public Money and Management*, Vol 28 (1), 57–60.

Seddon J (2005) 'Freedom from Command and Control' 2nd edition, Vanguard Press: Buckingham.

6

Organisational Change: Evaluating Systems Thinking in the UK Housing Sector – A Work in Progress

Joe Marshall

One approach to organisation and business improvement has been systems thinking (ST) and there is some recent evidence that systems thinking interventions in the service and housing sectors have made some significant improvements (Jackson, 2007 et al.; ODPM, 2005; McQuade, 2008; Seddon, 2007).

The purpose of this chapter is to report on the work in progress of a three year research project investigating the impact and sustainability of systems thinking on service performance in the UK housing sector. The research explores a number of ST interventions taking place in one of the largest housing associations in England, using an in-depth, real-time case study. The research proposes an evaluation framework and the key factors considered critical to the implementation, spread, and sustainability of systems thinking. Evaluation data collected during and beyond the formal ST intervention periods are presented and the chapter concludes that there are a number of quantitative and measurable organisational improvements and other benefits linked directly to the interventions.

Introduction

The list of approaches, methods, and models that purport to lead to organisation change and business improvement is a long one. Most approaches to organisational change to improve organisational performance have been evaluated in one form or another but often without convincing conclusions.

It is claimed that most organisational change programmes fail to achieve their objectives (Beer and Nhoria, 2000; Balogen and Hailey, 2004; Salaman, 2005). Armenakis and Harris (2009) cite a recent global survey of businesses that report that only one-third of organisational change efforts were considered successful by their leaders and the question is why?

The conceptual and prescriptive nature of change and improvement methods and models have been debated at some length and often critiqued as visionary and idealistic. It has been difficult to consistently confirm that particular methods and their implementation have led to improved organisational performance (Pettigrew et al. 1997). One of the problems is the confounding nature and number of organisational variables when looking for cause (a change intervention) and effect (organisational performance) relationships. Another problem has been the paucity of robust measures and evaluation criteria of successful change and its sustainability. Evaluation is difficult at the best of times and the process often contested, depending on whose data and opinions are sought, and even corrupted by the 'Macnamara Fallacy' (Butler, Scott and Edwards, 2003).

One approach to improve organisational and business performance has been the use of 'systems thinking'. Systems thinking interventions of one kind or another have been used in the UK public sector, health services, and elsewhere for some time and there is some recent evidence that systems thinking interventions in the UK housing sector have made some significant improvements (Jackson et al. 2007; ODPM, 2005; McQuade, 2008).

The particular systems thinking approach under investigation and reported in this research has been developed by Seddon (2003, 2008) and incorporates the work of systems thinkers (Deming 1982, 1994; Ackoff, 1999; Senge, 1990) and intervention theory (Argyris, 1993) in tandem with learning from Ohno's (1998) Toyota Production System (TPS) adapted for service organisations.

The purpose of this research is to examine and evaluate the systems thinking interventions taking place at a major social housing business in the UK in terms of the key factors that are believed to influence their overall success and sustainability (Buchanan et al. 2007).

This chapter begins by outlining the national and local context of the UK housing sector, and the case organisation. A review of key themes, concepts, and the literature that underpin organisational change and systems thinking interventions follows and an evaluation framework proposed that captures both the process and outcomes of systems thinking interventions over time. The research design and data collection methods are then described. The evaluation data is then presented with a discussion of the findings to date and finally report on the role and contribution of the work in progress.

National context

Nearly 4 million homes are owned and managed by professional social landlords in the UK; and although the number is significant the sector has become much smaller as a proportion of the overall housing market. The primary remit of the social housing sector is that of helping people who could not otherwise afford 'decent' housing, to do so. However, the scope for social landlords has now expanded, with most functioning at a local level with a social motivation that allows them to take a wider role in improving or protecting neighbourhood conditions, creating mixed-income communities, and even providing general advice to their tenants (for instance; debt prevention).

Housing associations are independent, not-for-profit landlords. They are regulated by government and reinvest surpluses into new housing and the neighbourhoods where they work. Housing associations provide homes for people in all kinds of circumstances. As well as affordable rented homes, associations offer supported housing to older and more vulnerable people, low-cost home ownership, key worker homes, and increasingly homes for sale on the open market.

The sector is highly regulated by the governments Audit Commission and the Tenant Services Authority (TSA) a new social housing regulator who will regulate, monitor, and analyse performance across the sector in England. A host of other regulators that work across the UK public services also impact on the work of housing associations.

Local context

Housing associations in the West Midlands region of the UK invested nearly £34 million between 2003 and 2008 on community infrastructure. They also run and support education, training, employment and local business start-up initiatives, childcare, after school and homework clubs, health, crime reduction, environmental, and many other projects. There are 175 housing associations in the West Midlands owning and managing nearly a quarter of a million homes, on behalf of nearly half a million people (NHF, 2007).

Central homes

Central Homes (CH) was created by a merger in 2006 of two leading housing associations in the Midlands and is now a major social business with assets of £1.1 billion, over 32,000 homes, employing more than 1600 staff. It is the largest housing and regeneration social business in the region and comprises four independent yet mutually supporting businesses. CH manages homes for rent, shared ownership, and houses for key workers and works with regeneration and community agencies and neighbourhood renewal programmes. CH is a major care and support agency providing homes and support for homeless people, those with disabilities and learning difficulties, older and younger people and operates and manages over 100 schemes.

In their corporate strategy for 2007–2012, the CH report on plans to shape the business over the coming years including new homes; responding to central governments 'sustainable communities' plans and local government's changing agenda; development of the community investment businesses of the group; propositions for CH growth and the preferred business model of 'Business Excellence'.

At a review of the business, in 2006 soon after the merger they declared that:

- Our landlord service was deemed as nothing special
- We have some high standards within our Care and Support services
- Our community development has a few examples of good practice but not coherent across the group
- Our property development needs additional work to compete with other organisations

Midway through 2007 the Business Transformation team asked Vanguard Consultants to undertake a 'scoping' exercise on 'General Needs-Income Management' and decided to proceed with a systems thinking intervention pilot which started in September 2007.

Organisational change and systems thinking

While the overall domain of this research is organisational change in the housing sector, the particular focus is that of a recent growing interest in systems thinking employed to bring about organisation change, improvement, and performance.

Systems thinking has a long history across many disciplines, schools, and scholars. Much of the early literature on the subject introduced the basic tenets and principles of systems thinking and these still apply and are used to the present day. The ideas of holism (as opposed to reductionism), connectivity, complexity, emergent properties, and feedback are reaffirmed from research publications, conferences, books, education programmes, and the media. Systems thinking literature often addresses problems of transformation, change, and redesign in organisations and the links and connection between systems thinking and the subject and study of organisational change is very apparent. A challenge in reviewing the literature and research is the sheer diversity and scope of both fields.

Advocates of systems thinking agree that changing the way that management and people in organisations think about problems systemically, organising resources, and people's contribution are major paradigm changes (Jackson, 2003). Much of the management and business literature on organisational change addresses the same issues and concerns that face systems thinking, theory and interventions. Systems theory as an organisational change perspective has long been advocated, for example in contingency models as Linstead, Fulop and Lilley (2004) remind us of this connection in their accounts of managing organisational change.

These authors and others (e.g. Collins, 1998; Harrison and Shirom, 1999; Senior, 2006; Hayes, 2007) cite the influence of open systems theory (von Bertalanffy, 1968), complex adaptive systems theory (Buckley, 1968) and chaos and complexity theory (Stacy et al. 2000; Fitzgerald, 2002; Burnes 2005) on organisational change theories.

Understanding organisations as closed or open systems has been the concern of management and organisational behaviour research and the systems interest in 'managing change' has been characterised as open systems (Iles and Sutherland, 2001). In a series of publications on organisational change for the UK NHS *Organisational Change – A Review for Healthcare Managers, Professionals and Researchers* a great deal of time is devoted to the explanation and contribution of systems thinking. The authors claim:

> Within the NHS the term *whole systems thinking* is now routinely used by managers and clinicians. (Iles and Sutherland, 2001, p. 17)

In many academic texts on organisational change, systems thinking theory, concepts and models appear as supporting work, for example viewing organisations as systems with properties of the whole rather than component parts. Organisations viewed as systems then should not be thought of as separate parts with chains of linear cause and effect relationships but as complex networks of interrelationships (Senge, 1990). Leading writers and thinkers on organisational change like Senge have appealed to both systems thinking and organisational change disciplines.

Seddon's approach to systems thinking

The particular approach to systems thinking studied in this research emanated from an investigation into the frequent failure of change programmes (Seddon, 2003, 2008). Seddon's philosophy and methodology is based upon the systems principles of viewing organisations and operations as systems (with a purpose always seen in terms of its customer, client, or citizen), holism, requisite variety, emergent properties (including unintended consequences), and the Toyota Production System in which success relies upon achieving economies of flow (not scale) and design against demand through the whole system, end-to-end.

A detailed account of the philosophy, theory and methodology is reported in Jackson et al. (2007) and Seddon and Caulkin (2007) and will not be restated here. A brief summary follows which highlights some key principles and methods of the approach and the conclusions drawn from the Office of the Deputy Prime Minister report on housing ODPM (2005).

Philosophy

Seddon describes his approach to systems thinking as different to others:

> managers [should] see systems thinking as a means to get knowledge ... and be able to act with prediction and confidence of improvement ... while all systems theorists agree a system is a sum of its parts and the parts must be managed as one, the Vanguard approach is unique in that it starts and ends with the work. (Seddon, 2003 p. 182)

Seddon cites Deming's (1982) argument that organisations should be managed as systems not functional hierarchies. Conventional management employs a command and control paradigm; is top-down and hierarchical, separated from the work, target and budget driven with an ethos of central control and reaction instead of learning and adaption. Using arbitrary targets drives behaviours to the detriment of the service to the customer Seddon argues throughout his critique of 'command and control' thinking. Seddon devotes much of his writing to the subject of how many public services in the UK including housing; social care and policing are driven by a belief in targets, incentives, and inspection. This way of thinking 'creates a de facto purpose i.e. meeting the targets which constrains method, because the work gets designed around the reporting requirements. When measures are derived from purpose [the customer's purpose] and those measures employed where the work is done, method is liberated' (Seddon, 2008 p. 82).

Systems thinking, is a way of designing, managing, and measuring the flow of work through the system as opposed to measuring and managing the work in functional activities. The effectiveness of the system, Seddon argues, should be measured in terms of customer purposes.

Understanding variety and variation in and on the system in manufacturing and service organisations is crucial. The 'variety' problem has been tackled by systems thinkers for some time, where the variety of demand placed upon the system is monitored and modelled thus reducing the variety of demand and making it more predictable. Customer value in service differs from manufacturing because of the variety of customer demand and Vanguards view of service organisations is that it should be front line staff who respond to the variety of customer demand. 'Demand in to service organisations can vary as greatly as the customers who co-create the service' (Seddon and Caulkin, 2007 p. 21). Service organisations, the authors argue, require frontline staff to be integral to the understanding and response to customer demands and the capacity to act autonomously is a case of 'learning' by those at the front end, not by centralised control.

> In service organisations it is the customer who sets the nominal value. In systems language the service organisation needs to be designed for customers to pull value from the organisation – to get exactly and only what they want, in the most expedient fashion. (Seddon, 2008 p. 69)

Methodology

The methodology 'Check- Plan-Do' is a variation of the 'Plan-Do-Study-Act/Plan-Do-Check-Act' model (Deming and Walton, 1992) used in many business improvement methods but one which starts with 'Check' rather than 'Plan': 'getting knowledge of the what and why of current performance as a system: nothing is assumed other than we almost certainly don't know what that performance is' (Seddon, 2008 p. 78).

'Check' is undertaken by the system intervention team (in-house staff directly familiar with the work) who will learn the approach by carrying it out. 'Check' examines:

1. the purpose of the service in customer terms
2. the types and frequencies of customer demand
3. how the system responds to demand
4. the flow of work; the value work and waste
5. the system conditions that create waste
6. the management thinking that governs current performance

Senge (1990, 1994) claims that systems thinking interventions alter thinking, behaviour, and results, while Seddon (2003, 2008) proposes a systemic relationship of 'thinking-system-performance' where 'thinking governs performance'. Both agree that it's the thinking that needs to change and that thinking should be systems thinking.

The findings from 'Check' are presented to the organisations stakeholders, including senior managers, customers, working partners, and those supporting the system. If agreement to proceed is granted then the next phases of 'Plan' and 'Do' can begin. Essentially this is a redesign of the system to achieve its new purpose from the customer's perspective. 'Plan' (or redesign) is the opportunity to design a system through continued experimentation and learning (action learning), that removes waste, achieves clean flow, and importantly aids managers and staff to change their thinking. These phases of the method could be slower than 'Check' with the overriding aim of 'doing it right rather than doing it quickly'. A number of methods are used by the systems team in the methodology above including; demand analysis, statistical process capability charts, flowcharts, system pictures, and logic pictures.

Conclusions

Seddon's distinctive approach to systems thinking has been examined and classified by Jackson et al. (2007) in his System of Systems Methodologies model (SOSM) and reported previously in some detail (ODPM, 2005; Jackson et al. 2007; Seddon and Caulkin, 2007; McQuade, 2008). The Evaluation Panel cited in the ODPM report stated that this systems thinking approach 'has the capacity to deliver significant efficiencies in service delivery. All the pilot organisations demonstrated potential annual six-figure gains from removing waste and redeploying resources more effectively' (Jackson et al. 2007 p. 8).

The study by Jackson et al. into the housing sector concludes that the systems thinking approach 'can be recommended as a useful methodology for bringing improvement to systems in the housing sector and, we believe, in many other areas of service delivery' (Jackson et al. 2007 p. 11).

The case studies cited in the above report have been carried out on Vanguard interventions in the housing sector and have focused on organisational performance and service improvements. However, they have focussed less on the factors that are thought to be critical to the spread and sustainability of systems thinking. A key purpose of this research is to examine these factors, to which we now turn.

Sustaining change and systems thinking

Considering the vast and diverse literature on organisational change the attention paid to the sustainability and spread of change is limited according to Buchanan et al. (2005, 2007). Armenakis and Bedeian (1999) reviewed the organisational change literature over a decade and identified four main themes; change substance, contextual issues, implementation processes, and criterion issues, i.e. outcomes. Sustainability was not considered as a criterion issue.

In a more recent and timely review, Armenakis has summarised his own research and practice on organisational change over the past 30 years (Armenakis and Harris 2009). The authors cite six 'signposts' of their research journey during this time:

1. change recipients motivation to change
2. change recipients active participation in the change effort
3. the importance of diagnosis
4. importance for creating readiness for the change
5. strategies for influencing change recipients motivation
6. the assessment of reactions to organisational change

The most significant signpost they argue is 'change recipient motives to support change efforts' and which 'therefore increases the likelihood of successful sustainable organisational change' (Armenakis and Harris, 2009 p. 127).

Armenakis and Harris go on to suggest that the final step of the change process is change adoption and institutionalisation. They suggest that the change recipients involvement and participation is central to increasing the likelihood of sustainable change. We return to this subject later when reporting on the changes taking place at Central Homes.

It would seem that until recently the issue of sustaining organisational change has not been significant in the literature and the same can be said of the sustainability and spread of systems thinking. Buchanan et al. (2005, 2007) provide an interesting case as to why this is, including the ambiguous nature of the concept of sustainability, the lack of research in the area which requires longitudinal study and devoted resources and that 'studying sustainability' is less interesting and stimulating than studying change.

Buchanan et al. (2005, 2007) propose a provisional model of key factors that affect the success, spread, and sustainability of various change initiatives. These factors, they suggest include: the external and internal organisational context; managerial and leadership style; scale of change and fit with organisation culture; power and influence; implementation methods and temporality; changes in thinking and shifts in mental models; and financial contribution. As yet Buchanan et al.'s model has not been tested, nor does it reveal the relative weightings and interaction effects among the factors. The gap in the research and literature on sustaining organisational change in general and systems thinking in particular provides an opportunity that this research will address.

Evaluating change and systems thinking

The evaluation of organisational change interventions is more complex than much of the literature suggests. First, accurate data depicting the organisations' response to a change process is difficult to collect and is open to corruption by the 'McNamara Fallacy':

> measuring what is easily measured... disregard what can't be easily measured... presume what can't be measured easily isn't important... and what can't be measured doesn't exist... this is suicide. (Handy, 1994 p. 219)

Second, the evaluative conclusions derived from the data are complex, high inference chains of reasoning based on implicit, taken for granted beliefs and values. Ontological and epistemological paradigms (i.e. the assumptions made about the nature of truth, human behaviour, reality, etc.) broadly determine the context for the conclusions of the evaluation inference even though they are rarely made explicit (Butler et al. 2003). The divergence of the interpretations of evaluation data, for example between managers, consultants, employees, customers and researchers, is an important issue in evaluation studies and it should be recognised that underlying assumptions exist. The problems that surround evaluation including the capture of individual cognitive processes and group interactions are heightened by the difficulties of the analysis of a dynamic and shifting organisational environment, both internal and external. One accepted view is that organisations promoting change want their staff to achieve changes at the affective, cognitive and behavioural level which will then impact upon the organisation in a positive way. The relationship between these levels, however, is thought to be difficult to discern and unravel. Studies have indicated that we know less about this relationship than professed (Allinger et al. 1997). One explanation proposed is the problem of data collection in the workplace (e.g. access to informant's and researcher effects) and the confounding issue of the 'McNamara Fallacy'. Another explanation is that the simplistic cause and effect relationship between individual learning and the organisational impact is flawed. This explanation suggests that a simple, linear and predictive relationship is not borne out empirically (Marshall et al. 2009) and is better explained by a non-linear, dynamic, emergent process that occurs between individuals 'thinking and doing' and the organisational response (Butler et al. 2003).

A deliberate aim of the evaluation process described in this chapter is to address two arguably conflicting paradigms of evaluation research. On the one hand to collect and examine quantitative data 'believed' to measure the impact of systems thinking on organisation performance and on the other hand to collect and examine qualitative data on the perceptions, thinking, and behaviour of individuals and groups before, during and after an intervention. Using a 'systems' epistemology to make sense of causality from various and perhaps conflicting perspectives embraces the view of Grint (2003) that a study of systems thinking such as this should refer to different accounts, various points of reference and competing descriptions of reality. This presents a fuller picture and understanding of the intervention process and outcomes and the factors that may lead to the sustainability of systems thinking.

According to Huz et al. (1997) four issues in the evaluation of systems thinking interventions should be considered:

1. Do organisations and norms/operating procedures change?
2. Does the quality of services delivered to customers improve?

3. Do individuals who participate in the interventions change the way they think?
4. Does the overall efficiency and effectiveness shift in a measurable way?

An issue that Huz et al. do not raise and is central to this research is: 'have the changes in the thinking, behaviour, service quality and organisation performance been sustained: and if not, why not?' This chapter particularly focuses on addressing this latter point in the evaluation of systems thinking.

Evaluation framework

An evaluation framework is now proposed which attempts to address the questions above and for a better understanding of the impact of systems thinking for a variety of the organisation's participants i.e. employees, customers, managers, consultants, and other stakeholders. To evaluate the interventions undertaken at Central Homes so far an effort is made here to explore the impact of interventions at different levels of analysis but also recognise that there is not a simple cause and effect relationship between changes in individual learning and thinking (levels 1 and 2) and organisational performance (level 3) (see table 6.1).

Research design and data collection

The research setting is Central Homes (CH) an organisation in the UK housing sector, a major social business managing 32,000 homes with a value in excess of £1.1billion. With a history of 80 years in social housing, CH is one of the largest housing associations in England. CH undertook their first systems thinking intervention (Income Management–General Needs) in September 2007. The second intervention (Income Management–Care and Support) began in May 2008 and the third intervention (Responsive Repairs) in March 2009. Access to CH was gained and exploratory field work and pilot studies undertaken for the research in early 2008.

The need for rigorous evaluative research is widely accepted in all disciplines, but recognised as a complex and difficult task. Much evaluation research is based on retrospective studies that are limited in tracking interventions over time. The design of the research described here tracks three interventions employing a retrospective, prospective and 'as-it-happens', real-time case study, over a period of more than two years which allows a fuller picture of the intervention process and outcomes. This offers a

Table 6.1   Levels and domains of evaluation

LEVEL 1	**Intervention sponsors and consultants – Assessment of intervention**
Domain 1	Business transformation consultant and advisor
Domain 2	Vanguard consultants
LEVEL 2	**Intervention participants and recipients – Perceptions of intervention and impact**
Domain 3	Intervention team members; perceptions of process and outcomes
Domain 4	Shifts in intervention teams' mental models, understanding of systems view
Domain 5	Changes in customer or client perceptions
Domain 6	Executive and senior managers, supporting functions; perception of process and outcomes
LEVEL 3	**Measurable systems change and business results**
Domain 7	Changes in business processes, policies, and procedures
Domain 8	Changes in customer or client service performance outcomes
Domain 9	Efficiencies, productivity improvements
Domain 10	Cashable savings

Source: Adapted from Huz et al. 1997.

'thick description' of the phenomenon in question (change processes) and is not an attempt to generalise to a wider population but as a case that arguably reflects a larger phenomena (Tsoukas, 2009; Yin, 2003).

Levels 1 and 2

Qualitative evaluation data has been and continues to be, collected and analysed at Levels 1 and 2 in the evaluation framework for a number of reasons. First, the 'process' of systems thinking interventions over time can be better understood by exploring the inter-subjective meanings that participants make of an intervention: before, during, and beyond the implementation period. This gives an indication of the motivation or otherwise of participants, to support and participate in organisational change efforts (Marshall and Olphert, 2008). Change participant motives are a key factor according to Armenakis and Harris (2009) in embedding new systems. Second, as Buchanan et al. (2007) advise, evidence linking an intervention and organisational performance requires an understanding of the events leading up to the current conditions and choice of intervention (path dependency); the range of effects leading to intervention outcomes (cumulative effects); and the particular combination of factors that lead to intervention outcomes (conjunctural causality). Third, the information gathered from intervention participants will indicate if shifts in behaviour, thinking, and their mental models have taken place, a critical aim of systems thinking (Huz, 1997; Seddon, 2003, 2008).

Evaluating qualitative data at Levels 1 and 2 (domains 1–6) is a critical aspect of this research and particularly for learning about key aspects of systems thinking methodology and its implementation. Much of the data is captured by an ethnographic approach 'as-it-happens' from the beginning of the second and the third intervention via participant observation of the intervention teams, unstructured and semi-structured interviews with other staff and from meetings and management engagement sessions. Data was also collected beyond the implementation period and retrospectively by focus groups; semi-structured interviews; learning/reflection sessions; and report minutes.

Level 3

Quantitative data evaluating at Level 3 of the framework was collected from company reports; archival documents; sponsors, meetings, and presentations. Interpreting performance data is a complex matter (as with all longitudinal studies) over periods beyond the implementation period or 'front-end' because of the inevitable changes in the environment along with changes introduced by the interventions. The nature of the confounding variables (e.g. new regulatory bodies, restructuring, new executive's/senior managers, and new policies and procedures) in any study looking for cause-and-effect relationships complicates the evaluation process.

Evaluation data

We now turn to the evaluation data and first examine evidence collected at Levels 1 and 2 in the evaluation framework.

Levels 1 and 2

The research to date has produced qualitative data from more than 60 individual participants involved over the three interventions undertaken at CH since September 2007. One note of caution about the data is the problem of informants telling you what informants believe you want you to hear. Researcher 'effects' are addressed to a certain extent here by the ethnographic approach, remaining in the field for more than a year to become 'saturated' with first-hand knowledge of the setting (Glaser and Strauss, 1967).

There is evidence from the data that the vast majority of the intervention team members and others, accepted and are positive about the value and effectiveness of systems thinking interventions, at different times and for different reasons:

A simple way of looking at what you have v. what you want. Then by following the ST rules i.e. demand, value, flow, and what matters and to whom. At all the schemes I have visited by the end all staff have agreed and commented that 'yes this will work'

Overall I think it was a required piece of work. I think the systems thinking processes are a much better way of analysing and designing working systems than have been employed previously by the organisation.

Nearly all participants found cognitive interest in the systems thinking principles and methodology and particularly during the 'Check' phase when the extent of failure demand, waste, and the system conditions (e.g. inspection, targets, budgets, IT systems) creating such poor service to customers and clients is revealed. Such a poor service, ironically at a time when key performance indicators and other corporate targets were being achieved e.g. acceptable customer satisfaction levels and repairs completed on time (as currently measured).

the level of failure was so high…I am really surprised at the failure…very poor service, I want to do something

Many participants reported that the learning and the change in their thinking and behaviour was a profoundly useful and fulfilling process both at the workplace but also away from the workplace.

Most critical is allowing staff who do the work on a day to day basis to have a large input on how they can improve the system. If you give people the opportunity to have ideas then it hopefully gives staff more motivation.

Managers reported that the 'management engagement' sessions had informed them of systems thinking concepts, the methodology, and their roles during the interventions and beyond. Most were enthusiastic about the likely improvements to service that could be made and their attitudes and motivation after the 'Check' presentation appeared to change demonstrably, expressing the need to 'do something' about improving performance.

I had no idea how long it takes to fix a repair... 1 day to 846 days... that's incredible

One theme that emerged from the data from some intervention team participants, sponsors and the consultants was a concern that some managers in the organisation have been less supportive than expected during the intervention process and to the principles and findings.

I think from the beginning of the intervention we did not get any involvement from senior managers or anyone asking how we were getting on etc. Maybe because people are too busy or that this is the process that has been decided and they leave staff to get on with it.

In terms of the process I feel that there should have been more emphasis on the senior management engagement sessions as I believe that they did not attend all of them and maybe did not understand the principles behind it and this is resulting in them slipping back to a command and control mentality.

Not all managers invited to attend the engagement sessions did so for various reasons and did not take part in the data collection tasks planned between engagement sessions. This supports the comments made by the intervention team members on the 'lack of buy-in' as discussed above.

Overall, consultants and sponsors expressed very positive views on the progress and success of interventions, as might be expected. However, they often discussed an intervention in terms of the disappointing levels of engagement from some senior managers and executives. Conversations often concerned their perception on the 'lack of leadership at the top' for the successful implementation of systems thinking in the organisation.

The senior team aren't engaged ... some have attended engagement sessions but still aren't convinced. Senior managers are still not involved, they need to mobilise HR, Finance, IT to support systems thinking ...

Executives and senior managers including those from partner and contactor firms reported their feelings of surprise, incredulity, and even shame on the 'current performance' data (i.e. levels of failure demand and waste) at the 'Check' presentations.

...it highlights our ignorance...An average of 88 days to do a repair! This is disgraceful.

The cost of waste is around £1.5 million and that's out of all our pockets...

The evaluation to date which explores Levels 1 and 2 (domains 1–6) presents a partial but mainly positive view of the interventions from a wide range of organisation participants and other stakeholders. The data collected thus far helps us to explore and support key factors suggested in the literature as critical to the overall success and sustainability of new systems.

Level 3

The evaluation data collected at Level 3 and summarised in Tables 6.2. and 6.3 suggests that there have been a number of measurable and quantitative improvements as a result of the two systems thinking interventions undertaken at CH to date. Evaluation of the domains at Level 3 are witness to what the transformation team believe are 'excellent results' and they note the 'significant potential for financial improvement'.

Analysis and discussion

Buchanan et al. and other scholars from the 'process' explanation of organisational change suggest that sustainability and spread of change and new methods are explained by a model of conjunctural causation, i.e. the occurrence of particular configurations of 'fateful' factors. These factors and conditions are now discussed with regard to the findings to date under three broad analytical categories of substance (content), context and process of change.

Intervention/change substance

The *Perceived Benefits* arising from change interventions is considered a key factor of success. The intervention method evident in this study achieves strong credibility in terms of the extent of the 'current performance' problems found in all three of the interventions. The benefits to customers, staff and the business in resolving these problems by systems thinking were communicated to senior managers and other stakeholders. The hard evidence collected by the intervention teams make a resounding impression on their audience to support and act on the evidence. The data on improvements to the business so

Table 6.2 Intervention: Income management – general needs

Summary of results	Evaluation domain
94 new lettings surveys completed using new redesigned flow. 95 percent scored the service 7/10 or above: 67 percent scored 10/10	5 and 7
Affordability assessments now undertaken for new customers (focus on financial inclusion and sustaining tenancy).	7
Rent payment now made at sign up (preventing debt accrual).	7 and 8
Income calls now directly to income management team (one stop service)	8
Rent collection now 100 percent (arrears reducing significantly); bad debt favourable variance of £739k.	10
General needs rent collection sustained at over 100 percent (4 percent increase from 07/08 to £4.32m).	9
Cashable interest saving of £117k.	10
Arrears increased 07/08 (£1.5m); same period 08/09 increased by £500k.	9
Rent payment team now represent at Court. Solicitors savings £17k.	10
Court applications now actioned online. Saving £9k.	10
Changes to income mgt. processes now made by operational staff	7
Days taken from tenancy commencement date (TCD) to first payment posted on account fallen from an average of 44 days to 3.6 days.	9
Home visits increased from 593 to 1341 (closer relationship with client/customer).	8
Cash applications team now located with income management team.	7
Joint working practices developed e.g. coordination of evictions, abandonments, abandoned vehicles, access for gas servicing, and co-joined legal application.	7

Source: Extracted from 'Central Homes' reports until October 2008.

far, shown in Tables 6.2 and 6.3 indicate the benefits clearly enough and communicating this to staff at all levels should encourage the spread of systems thinking. If Armenakis and Harris (2009) are correct in their claim that 'recipient beliefs and motives to support change efforts is the most significant signpost' then change 'recipients' at CH would appear to have the 'motives' to support and embrace systems thinking.

Table 6.3   Intervention: Income management – care and support and shared ownership

Summary of results	Evaluation domain
New specialised C & S income management team established. Ownership of scheme and income management will continue to reside at scheme level.	7
ICT the Kirona Solution utilised for 'General Needs' adapted for C & S. Access to Northgate to be restricted for scheme staff (risk management).	7
Welfare benefits and money advice training for all scheme staff. Specialist legal training for managers and C & S income management team.	7
Ongoing communication to inform and engage all affected staff. Briefings for senior and operational managers (roles in supporting new ways of working). Core brief article and presentation delivered to all staff. Currently working on plan to communicate results achieved to date to key stakeholders.	7
Arrears reduced from £85.3k to £39.4k in eighteen weeks ('Homeless Youth'). Arrears reduced from £67.7k to £34.6k in eighteen weeks ('Homeless'). £7k reduction in arrears at 'Zambesi' scheme (first pilot after one week).	9
£463k not reconciled income recovered. All SP account changes to date resulting in overall reduction of £1.78m. £308k reclaimed from suspense account.	9

Source: Extracted from 'Central Homes' reports until December 2008.

The *Integration* of systems improvement is another factor. System changes in isolation and given priority or implemented in a way that compromises other services are potential threats to sustainability. The interventions selected and undertaken so far at CH have and will affect other departments of the business (e.g. Finance, HR and IT) in terms of the design and delivery these services presently offer. Their roles and functions will need to change to support systems thinking. This is an issue that systems thinking sponsors and senior management must address if the perceptions held by those who believe their professional judgements, roles, and service is compromised and even threatened in any way. Managers and staff from these departments were often those that expressed their reservations about the value of systems thinking.

Intervention/change context

Regardless of the change methods, the culture of the organisation is thought to be important in successful change. Conflicting priorities, multiple agencies, readiness, resources, current performance targets, and the history of change efforts, will all influence the momentum and sustainability of change. The performance target culture at CH (KPI's call centre targets, tradesmen productivity, and arrears targets) and other 'system conditions' (schedule of rates codes, contractual agreements, ICT interfaces) like many public services in the UK is very evident. Systems thinking focused on 'customer purpose' is often in conflict with the targets in use. A key issue for the success of systems thinking in the longer term is senior management's attention focussed on the regulatory target regimes and other priorities of the business e.g. post merger legacy problems at CH, which divert attention away from systems thinking. The conflicting priorities of targets versus systems thinking were expressed by all participants, many times and feelings were that this conflict is a major barrier. Balancing priorities for executives and senior managers is a complex matter; it is a bold CEO who ignores the regulators. There is evidence that change initiatives are less likely to succeed if not linked to organisational priorities and when viewed in isolation, rather than complementary (Buchanan et al. 2005).

Intervention/change process

Widespread early *Engagement of Staff* at individual, group, and discipline levels has been identified as a key factor in facilitating change. The approach and methodology of systems thinking does indeed give the intervention team and those directly involved ownership and responsibility of designing new systems. The overwhelming feedback by those undertaking 'Check-Plan-Do' was the value in participating in systems redesign. What was apparent was the limited participation of senior managers in communicating and selling the message. For example the training and development department has not initiated, or been asked to initiate training, development, or familiarisation events on systems thinking. This appears to be an opportunity not taken to address the *engagement* factor.

Key roles, power and influence

The role of CEO and senior management is a 'crucial and pivotal' factor in the sustainability and spread of systems thinking. We discussed earlier some concerns expressed by consultants, sponsors, and intervention teams on the limited involvement of some senior managers. As key gate-keepers to embedding change and the survival of systems thinking at CH a lack of support at this level could be critical. Where their active involvement or interest is absent and/or not conveying clear messages and priorities, systems thinking is unlikely to stick.

The role of the 'business transformation team' as sponsors of systems thinking is also critical. Their enthusiasm, persistence, tenacity, and credibility in terms of the project management of interventions, working closely with staff, identifying movers and shakers, and influencing peers is important. Consultants expressed feelings, during all three of the interventions, not about the technical knowledge of the internal sponsors, but on the need for them 'to get things moving more' when for example, conflicts appeared between different professional and functional groups. 'Care and Support' schemes at CH for example, employ scheme managers and housing officers who often express the dilemma in their roles of social worker/carer and that of debt collector. Sponsors of systems thinking are 'opinion leaders' whose success depends on social networks rather than formal authority positions and are critical in pluralistic organisational settings like CH (not-for-profit, multiple stakeholders, clients and tenants, care agencies, and so on). For systems thinking to embed and continue at CH the task of seeking and adopting more opinion leaders is important. As the interventions at CH progress more opinion leaders may emerge but more may need to be done.

Temporal issues

The time to focus and dedicate time away from operational demands is another key factor. The systems thinking methodology under review here is predicated on the dedication of the intervention team to experiment and learn in 'Plan' (redesign) phase. Systems thinking is beyond a short-term quick fix and needs time to embed. To make a 'cultural shift' to systems thinking will demand a long timeframe.

The 'fateful' factors discussed above are helpful in understanding some of the conditions for sustaining or the decay of systems thinking but what is not clear is the relative importance or weighting of the factors. First, do policies, systems, and structures have more influence than the commitment of senior managers to systems thinking? Second, what are the interactions between the factors? Will the redesign of 'Responsive Repairs' threaten to upset the status quo of power held by managers at CH and contracting firms? Third, the changing context will almost certainly influence the weighting and interaction effects. New regulatory standards may promote the value of 'customer purpose' measures and hence systems thinking. Changes of staff at the executive level or the desire to grow CH by acquisition may divert the desire to extend systems thinking throughout the organisation. To date and nearly two years on there is some evidence of embedding systems thinking at CH but time will tell and research will continue to track progress and report on any 'cultural shift' at a later date.

Conclusions

An important contribution of this research is the evaluation framework designed to explore the impact, and likelihood of the spread and sustainability of systems thinking. The framework is valuable in terms of the levels and domains of evaluation at different points in time that expose competing interpretations of a systems thinking intervention but more important reveal the underlying factors and conditions that are likely to influence a change to systems thinking, continued benefits to the organisation, customers and other stakeholders in the longer term.

The evaluation presented in this chapter is a work in progress. The results to date present a partial but favourable picture of systems thinking for a variety of different stakeholders and particularly the performance of service delivery in terms of 'customer purpose'. The ultimate aim of the research, however, is to explore the subsequent events beyond the short and medium-term and make an evaluation over a longer timeframe which will help identify the 'fateful factors' that will or will not lead to a systems thinking organisation.

There is a body of knowledge that suggests 'evaluation begins in the design phase of change programmes not after a programme has ended' (Laird, 1985 p. 267) and arguably some systems thinking approaches 'design in' evaluation from the start. However, the systems thinking logic of Seddon's approach makes a strong case which declines to plan and predict performance improvement, cost benefits and deliverables on the grounds that we almost certainly don't know what 'current performance' is. What accounts of Seddon's approach do reveal is that improvements on 'current performance' for the customer and the organisation are almost a certainty.

References

Ackoff, R. (1999) *Ackoff's Best*. New York and Chichester: John Wiley.

Allinger, G. M., Tannenbaum, S. I., Bennett, W., Traver, H. and Shotland, A. (1997) 'A Meta-analysis of the Relations Among Training Criteria'. *Personnel Psychology*, 50, 341–360.

Argyris, C. (1993) *Knowledge for Action. A Guide to Overcoming Barriers to Organisational Change.* San Francisco: Jossey Bass.

Armenakis, A. A. and Bedeian, A. (1999) 'Organizational Change: A Review of Theory and Research in the 1990s.' *Journal of Management*, 25(3), 293–315.

Armenakis, A. A. and Harris, S. G. (2009) 'Reflections: Our Journey in Organisational Change Research and Practice.' *Journal of Change Management*,.9(2), 127–142.

Balogun, J. and Hailey, V. H. (2004) *Exploring Strategic Change.* 2nd Edition. Pearson Education Limited: Harlow.

Beer, M. and Nohria, N. (2000) 'Cracking the Code of Change'. *Harvard Business Review*, May–June, 78(3), 133–141.

Buchanan, D., Fitzgerald, L. and Ketley, D. (Eds.) (2007) *The Sustainability and Spread of Organisational Change.* London: Routledge.

Buchanan, D., Fitzgerald, L., Ketley, D., Gollop, R., Jones, J., Saint Lamont, S., Neath, A and Whitby, E. (2005) 'No Going Back: A Review of the Literature on Sustaining Organisational Change'. *International Journal of Management Reviews*, 7(3), 189–205.

Buckley, W. F. (1968) *Modern Systems Research for the Behavioural Scientist.* Chicago: Aldine.

Burnes, B. (2005) 'Complexity Theories and Organisational Change' *International Journal of Management Reviews*,7(2), 73–90.

Butler, J., Scott, F. and Edwards, J. (2003) 'Evaluating Organisational Change: The Role of Ontology and Epistemology'. *Tamara: Journal of Critical, Postmodern Organisation Science*, 2(4), 55–67.

Collins, D. (1998) *Organisational Change: Sociological Perspectives.* London: Routledge.

Deming, W. E. (1982) *Out of the Crisis.* Cambridge: Cambridge University Press.

Deming, W. E. (1994) *The New Economics: For Industry, Government, Education.* 2nd Edition Massachusetts: MIT Press.

Fitzgerald, L. (2002) 'Chaos: The Lens that Transcends'. *Journal of Organisational Change Management*, 15(4), 339–358.

Glaser, B. G. and Strauss, A. L. (1967) *The Discovery of Grounded Theory: Strategies for Qualitative Research.* Aldine Publishing Company: Chicago.

Grint, K. (2003) 'Problems, Problems, Problems: the Social Construction of Leadership'. *Human Relations.* 5(11), 1467–1494.

Handy, C. (1994) *The Empty Raincoat: Making Sense of the Future.* Hutchinson: London.

Harrison, M. I. and Shirom, A. (1999) *Organizational Diagnosis and Assessment; Bridging Theory and Practice.* Sage Publications: London.

Hayes, J. (2007) *The Theory and Practice of Change Management.* Palgrave Macmillan: Basingstoke.

Huz, S., Anderson, D. F., Richardson, G. P. and Boothroyd, R. (1997) 'A Framework for Evaluating Systems Thinking Interventions: An Experimental Approach to Mental Health System Change'. *System Dynamics Review*, 13 (2), 149–169.

Isles, V. and Sutherland, K. (2001) *Organizational Change – A Review for Health Care Managers, Professionals and Researchers.* NCCSDO: London.

Jackson, M.C., Johnston, N. and Seddon, J. (2007) 'Evaluating Systems Thinking in Housing.' *Journal of the Operational Research Society*, Nov, 1–12.

Laird, D. (1985) *Approaches to Training and Development (2nd ed revised)*. Reading MA. Addison Wesley.

Linstead, S., Fulop, L. and Lilley, S. (2004) *Management and Organization: A Critical Text*. Palgrave: Palgrave.

Marshall, J. M., Buxton, S. and Smith, S., (2009) 'Learning Organizations and Organizational Learning: What Have We Learned?' *International Journal of Knowledge, Culture and Change Management*, 8(5), 61–72.

Marshall, J. M. and Olphert, A. (2008) 'Organizational Change in the National Health Service: lessons from the staff.' *Strategic Change*.17, 251–267.

McQuade, D. (2008) 'New Development: Leading Lean Action to Transform Housing Services.' *Public Money and Management*, February, 57–60.

National Housing Federation (2007) *Home Truths*. Birmingham. NHF West Midlands

ODPM (2005) *A Systematic Approach to Service Improvement: Evaluating Systems Thinking in Housing*. ODPM Publications: Wetherby.

Ohno, T. (1988) *The Toyota Production System*. Productivity Press: Portland, Oregon.

Pettigrew, A. (1997) *The Double Hurdles for Management Research. In T. Clarke (Ed), Advancement in Organisational Behaviour: Essays in Honour of D. S. Pugh*. 277–296 London: Dartmouth Press

Salaman, J. G. (2005) *Bureaucracy and Beyond: Managers and Leaders in the Post-Bureaucratic Organisation in P. du Gay* (ed) *The Values of Bureaucracy*. Oxford University Press: Oxford.

Seddon, J. (2003) *Freedom from Command and Control*. Vanguard Press: Buckingham.

Seddon, J. and Caulkin S. (2007) 'Systems Thinking, Lean Production and Action Learning'.

Action Learning: Research and Practice, 4(1), April 2007, 9–24.

Seddon, J. (2008) *Systems Thinking in the Public Sector: the Failure of the Reform Regime and a Manifesto for a Better Way*. Triarchy Press: Axminster.

Senge, P. (1990) *The Fifth Discipline: the Art and Practice of the Learning Organisation*. Random House: New York.

Senior, B. and Fleming, J. (2006) *Organisational Change*. Pearson Education Ltd: Harlow.

Stacey, R., Griffin, D. and Shaw, P. (2000) *Complexity and Management: Fad or Radical Challenge to Systems Thinking*. Routledge: London.

Tsoukas, H. (2009) *Craving for Generality and Small-N-Studies: A Wittgensteinian Approach towards the Epistemology of the Particular in Organisation and Management Studies in Buchanan and Bryman* (eds) *The Sage Handbook of Organisational Research Methods*. Sage Publications: London.

Von Bertalanffy, L. (1968) *General Systems Theory: Foundations, Development, Applications*. George Braziller: New York.

Yin, R. K. (2003) *Case Study Research: Design and Methods*. Sage Publications: Thousand Oaks, CA.

7

Rethinking for Radical Improvement in the Delivery of Housing Services: An Overview of the Application of Systems Thinking in the Housing Sector

Donna Samuel and Barry Evans

In 1990, a book called *The Machine That Changed the World* shook the automotive industry by showing how far ahead certain Japanese car makers were in manufacturing, design, and supply chain capability using lean production techniques pioneered by Toyota over the last 50 years. The term lean was initially used then to coin or describe Toyota's Production System (TPS), more recently it is used to describe Toyota's entire management system. However, since that publication the sphere of influence of lean has been vast and ubiquitous. Originating in automotive manufacturing, lean has spread into general manufacturing, into the service sectors such as retailing and, more recently, into public and third sector organisations. Moreover, many have argued that the term lean is highly inappropriate for describing the depth of approaches deployed in Toyota.

A major exponent of TPS in the service sector is Professor John Seddon. Seddon argues that systems thinking 'underpins' lean management and that TPS is probably the most highly developed, best articulated and successful example of systems thinking applied to a business organisation in the world (Seddon and Caulkin, 2007). Over time, Seddon has gathered a body of evidence to demonstrate the efficacy of his way of applying systems thinking in a diverse range of service organisations (both private and public sector), including a number of examples in housing. Housing is a sector at the forefront of systems thinking and one where a body of empirical evidence is emerging. This chapter adds to the emerging body of empirical evidence.

The purpose of this chapter is not to explain Seddon's methods, since this is well-documented elsewhere (Seddon, 2005; OPDM, 2005; Seddon, 2008), but to examine its impact and effectiveness through a review of the small but growing body of literature and through the presentation of case study data. The authors will draw some tentative conclusions as to the potential benefits for the housing sector more broadly.

The title of this chapter is deliberately provocative since systems thinking[1] in the housing sector could mean dramatic improvement in service delivery for those organisations brave enough to experiment.

Introduction

The influence of lean in the private sector has been widespread and profound and is well-documented in the literature (for recent literature reviews see Papadopoulou and Ozbayrak, 2005; Bhasin and Burcher, 2006). The term lean was first coined by Krafcik (1988) and later popularised by Womack et al. (Womack et al. 1990; Womack and Jones, 1996, 2005) to capture the essence of Toyota's Production System (TPS). The word lean was selected to reflect the far less resource-hungry TPS as compared with the typical production systems in the West. The managerial characteristics that result in this production system are many and diverse but revolve around a systematic elimination of waste in all its forms together with a focus on continuous improvement (Papadopoulou and Ozbayrak, 2005). The definitions provided in the literature are unclear. Some authors have commented that the accumulated body of research on lean is primarily anecdotal rather than enlightening (Spenser and Guide, 1995). Others argue that the lack of clear definitions is explained by the fact that lean has evolved out of experimentation and is still in development through experimentation (Papadopoulou and Ozbayrak, 2005). Lean, then, is both a nebulous and an evolving concept, whilst originally conceived as a counter-intuitive alternative to traditional manufacturing (Womack et al. 1990; Shingo, 1989; Krafcik, 1988), it is now presented as a new paradigm for operations (Katayama and Bennet, 1996; Bhasin and Burcher, 2006; Bartezzaghi, 1999).

Lean remains one of the most influential new paradigms in manufacturing and has expanded well beyond its original application on the shop floor of vehicle manufacturers (Hines et al. 2004). Since its introduction into mainstream management thinking in the late 1980s, lean or lean manufacturing continues to attract considerable attention, from both the practitioner and academic communities.

Furthermore, the lean circle of influence has spread from its origins in car making and manufacturing into the service (Swank, 2003) and public sectors (Radnor and Boden, 2008; Radnor et al. 2006; Bagley and Lewis, 2008; Papadopoulos and Merali, 2008; Seddon, 2008; Seddon and Brand, 2008) including the health sector (Bushell et al. 2007; Beata et al. 2007).

Background

A major exponent of TPS in the service sector is Professor John Seddon; however, he has stopped using the term 'lean' due to some common misconceptions around the term which associate it with the application of a set of tools rather than thinking. Seddon argues that systems thinking underpins lean and that TPS is a striking example of systems thinking applied to a business organisation (Seddon and Caulkin, 2007). Notwithstanding its difficulties in recent months, Toyota has long been one of the most profitable automotive companies in the world and has achieved market dominance through an alternative way of thinking about the design and management of work and 50 years of applied continuous learning. Seddon's approach to the translation of TPS for non-traditional lean environments such as services and public sector has been well documented elsewhere (Seddon, 2005; Seddon, 2008; Seddon and Caulkin, 2007; Seddon and Brand, 2008; Advice UK, 2008; Zokaei et al. 2010).

Seddon (2008) makes a powerful argument that the public sector is littered with examples of poor service often as a result of the imposition of targets and inspection driving the wrong behaviour and that by taking a systemic approach to service design, dramatic service improvement can be achieved quickly. Built on a long tradition of systems theory (see Flood, 1999 for a brief summary), essentially, systems thinking offers a way forward for decision makers faced with the failure of mechanistic and reductionist thinking when confronted with complex, real-world problems, set in social systems (Chapman, 2004; Jackson, 2005). Jackson et al. (2008) evaluate Seddon's lean systems approach (term used by Jackson et al. to describe the approach) using a critical systems thinking device known as the 'systems of systems methodology' (SOSM), (Jackson, 2000, 2003). SOSM was first devised by Jackson and Keys in 1984 and is the most cited way of classifying systems methodologies (Jackson et al. 2008). Their evaluation finds that Seddon's lean systems approach provides a well-specified methodology embodying many aspects of systems thinking. Criticisms of the approach include: its failure to accommodate a variety of possible purposes; and, the risk of sub-optimisation (optimizing one subsystem without reference to the other parts or levels of the system).

Comprehensive descriptions of the lean systems approach can be found elsewhere (Seddon, 2005, 2008), the list that follows highlights the key characteristics:

1. The approach begins with a consensus review of the true purpose of an organisation or service. The simple articulation of a unifying purpose is powerful in two key ways: first, it reveals the fact that the service is often built around de facto purposes derived from arbitrary targets; second, it provides a yardstick or sanity check for potential redesign ideas.
2. The approach includes a systematic analysis of demand facing the service organisation. Demand is designated as value demand (the demand for which the service is intended to meet) and failure or preventable demand (the additional demand or work generated as a result of service failure). The articulation of value and failure demand enables the service to be redesigned and reveals the extent of wasted organisational capacity.
3. The approach involves studying the flow of work, or how everything works end-to-end from the customers' point of view. The true end-to-end time it takes to provide service, when expressed as a capability (time-series) chart, will expose the predictability of performance and its variation. Identifying the causes of variation leads to an understanding of the need to remove the causes in order to reduce time and thus to improve the service. Critically it uses time-series analysis to expose 'common cause' variation which is to be expected in any system and should be 'tolerated' and 'special cause' variation which warrants attention if and when they reoccur. Second, service improvement is measured by narrowing the range between upper and lower control limits of 'common cause' variation.
4. The service is redesigned against customer demand. The redesigned service incorporates customer regulation, workers responding to customer need and managers making it easier for workers to do.

Research methodology

Seddon has been developing this lean systems approach for some time while simultaneously gathering evidence for both its efficacy and its impact (Seddon, 2005, 2008). This chapter focuses in on the work that has been conducted so far in the housing sector. The housing sector is one in which there is some information currently in the public domain and one which has been particularly proactive in experimenting with the lean systems approach. The research aims to address the following question:

How effective has experimenting with the lean service approach been in the housing sector in the UK?

The research methodology incorporates two elements: a review of the literature currently available on lean service approach in the housing sector and illustrative cases studies drawn from interviews with two housing associations in Wales currently experimenting with the lean service approach. Hakim (1987) argues that case studies are the most flexible of all research designs. Yin defines a case study as:

An empirical enquiry that investigates a contemporary phenomenon within its real-life context when the boundaries between phenomenon and context are not clearly evident, and where multiple sources of evidence are used. (Yin, 1989, p. 23)

The case studies presented here are exploratory (Yin, 2003) and instrumental (Stake, 2000). They have been drawn up as a result of in-depth discussions with individuals in the two organisations who have been closely involved with the lean systems approach. Discussions were not taped. The information presented here represents the author's interpretations from notes produced during these discussions. However, key individuals in both cases have approved the interpretation presented and are happy for their and the names of their organisations to appear in this chapter.

Lean systems in the housing sector: The work to date

In 2004 the Gershon inquiry into public sector efficiency made its report on releasing resources to the frontline. Gershon stated that efficiencies could be achieved by delivering the same or better outcomes with the same or fewer resources. The response in the public sector has seen the introduction of an array of efficiency targets which have frequently resulted in damaging unintended consequences (Seddon, 2008).

In 2005, in response to the so-called Gershon review, the then Office for the Deputy Prime Minister (OPDM, 2005) produced a report on research they had commissioned into the application of systems thinking in housing. The research was based on three pilot organisations and covered a range of services. The findings of the report were that the pilots indicated that systems thinking has the potential to deliver major efficiencies in service delivery: 'The work undertaken in all three pilots demonstrates cashable and non-cashable efficiency gains and significant service improvements' (OPDM, 2005).

A year later, a follow-up report was produced by Northern Housing Consortium (NHC, 2006) with the sole purpose of investigating the sustainability of the approach in the three pilots. These two documents, then, report the improvements in three organisations where systems thinking interventions had taken place at two points in time, i.e. during (or right after) the intervention and a year or so later. Table 7.1 summarises the improvement the pilot organisations realised at these two points in time.

The follow-up report states that 'systems thinking has become embedded in two of the three pilot organisations', with the reason for the remission of one pilot explained by organisational merger. The report concludes that early results show the potential for substantial gains. The same piece of work has since been presented in the academic literature (Jackson, Johnston and Seddon, 2008).

Wales has recently seen the publication of a report on affordable housing commissioned by the Deputy Minister for Housing, the so-called Essex review (June 2008), in response to the commitments made in *One Wales*, the policy programme for the Welsh Assembly Government (WAG) for the period until May 2011. Amongst the many recommendations is one which calls for the WAG to 'play a much more prominent role in the overseeing of the health and performance of housing associations in Wales... inspection needs to be a core part of the new regulatory regime' (Essex, 2008).

This suggestion that the response to system failure should be more inspection is concerning. It is hoped that this chapter explains why and suggests that there may be another more effective course of action in the systems thinking approach. The following case studies explain how systems thinking can radically improve the delivery of housing services.

Case study findings

Case study 1: Coastal housing association

Coastal Housing Association (formerly Swansea HA) is led by systems thinker Tim Blanch. Tim has always had a healthy disregard for government targets but failed to truly appreciate why until he heard Seddon

Table 7.1   Improvement in pilot organisations during and after interventions

Pilot organisation	Improvement in 2005 during intervention	Improvement in 2006 post intervention	Explanation/caveat and comments
THG (repairs)	End-to-end time reduced from 46 to 6 days Potential 6 figure efficiency gains	End-to-end time remains at 6 days Efficiency gain calculated at £115,000	Improvement sustained Anticipated level of gains not yet reached the forward look expectations of the original report
LSEH (re-housing)	Void re-let time reduced from 50 to 25 days Number of voids fell from 240 to 150	Void re-letting time risen to 34 days Number of voids found to be 116 leading to efficiencies on void loss of over £30,000	This is as a result of letting properties that had been empty for some time, increasing the overall average
PCC (rent collection and debt recovery)	Rent collection for new tenants reduced from 34 to 20 days New tenants falling into arrears reduced from 43 percent to 18 percent	Unclear Risen back to previous levels	The improvements in the rent collection and debt recovery subsystems could have some negative impact on other subsystems (namely voids)

speak some years ago and embarked on a learning journey he is still enjoying. Tim admits that budget control in his organisation is secondary to getting the job done and insists that the organisation is better off for it. Decision-making in the organisation has shifted to front-end staff and the result is empowered employees and satisfied customers.

Coastal started on their systems thinking experiment some five years ago, selecting lettings as a initial starting place for a variety of reasons (including little third party involvement, few external regulatory KPIs and the slowness, complexity, and wastefulness of the current resource-hungry process). The aim was to fundamentally review the applications process from the customer's perspective. The work was conducted by a mixed team of individuals including both frontline worker and managers. Reviewing the process involved listening to customers and finding out what they were asking for. The work quickly revealed that customers typically want two questions answered: Am I eligible for a letting? If so, can I be offered a suitable property? A detailed review of the current process produced some revealing (though probably not untypical) findings: up to six members of staff were involved in processing each application; each application involved the organisation sending out at least two letters (whether successful or not); 80 percent of applications were awarded on medical points; complexity was even greater than had been thought especially with regard to disabled applicants, those requiring support and referrals from other partner agencies.

The redesign of the process involved making the application process telephone based. Applicants receive an initial decision on their eligibility in a single phone call lasting about 10 minutes. During the phone call, successful applicants are offered a home visit at a time that suits them. The time elapsed from an application and home visit has been reduced from 47 to just 6 days. Satisfaction with the application system has increased with fewer applicants making complaints in spite of the fact that increasing demand means that a lower proportion of applications are successful. The redesigned process has resulted in significant savings in stationery and printing .

Coastal Housing has conducted similar intervention and redesigns on their maintenance, transfers, and rental incomes processes. The first three interventions involved the use of external consultants but more recent ones have been conducted entirely independently using the expertise they have developed internally.

During their various interventions, Coastal observed the following:

- Demand analysis – actually 60 percent failure demand, 40 percent value demand (fairly normal for each intervention at the 'Check' stage)
- 50 percent of people that they wrote to with appointments did not confirm, means that follow-up letters were then sent. Of the ones who did confirm only 75 percent kept the appointment!
- The length of time (end-to-end process time) was enormous (207 day median) with some cases up to 750 days. Transfers were worse with a median of 587 days.
- Their processes involved lots of failure demand, batching, checking, audit trails, and acceptance of partial information.

Redesigning the basic process around what the customer actually wants releases both cash and capacity for better use. Coastal has such confidence in their approach that during the Wales Audit Office (WAO) visit in 2006, they decided not to pretend that they had the data required for the statutory targets. Instead they invited the commission to see the work they were doing and to conduct focus groups. The WAO's findings were positive and are in the public domain (www.wao.gov.uk).

Case study 2: Charter housing association

Charter Housing Association is also fortunate to have an innovative leader in Kathryn Edwards. Kathryn also noticed that against traditional KPIs her organisation was achieving 98 percent completion within timescales being reported and yet at the same time complaints were increasing. This anomaly acted as a catalyst for action.

Charter started experimenting with systems thinking just over two years ago. So far they have done two interventions: the first on maintenance and repairs, the second on the more complex lettings and allocations process. The first two involved consultants but they are confident that they have acquired enough internal expertise to 'go it alone' for the next two interventions planned for this year.

The senior management team were involved at the outset and attended a course one day a week over six weeks. The purpose of this training was that they understood the nature, context, and purpose of the work that was to follow.

The first intervention, on repairs, took 10 weeks and involved a team of four staff (again a mix of frontline and management) and consultants were employed for 30 days. The approach taken was to establish what matters to customers, to revisit the purpose of the process, and finally to redesign the process around that purpose and their findings.

The first five weeks, the 'Check' stage involved listening to incoming calls. Many of the calls (49 percent) were failure demand (e.g. 'I reported a fault but have not received a repair appointment, the contractor did not turn up, the contractor turned up but it is still not fixed'). The check stage revealed that what actually matters to customers is: being kept informed, appointments being made and kept, faults fixed first time, total time taken to repair, repair remains fixed, and good quality workmanship. End-to-end time was found to be up to 346 days with a mean of 78 days. In general, the team learned that the repairs system was driven by targets as opposed to the customer. The result was staff time spent on waste producing information and driving processes that were of no benefit to the customers.

Their findings led the team to redesign the process and develop measures around their revised purpose. The purpose of the process was redefined as: 'do the right repair right first time at a time that suits the customer'.

The team presented their proposed redesign to the senior management team. Kathryn recalls the senior management team's concerns that they would be inundated with repairs and would not be able to cope with the surge in demand. However, this was not the case. In fact, prior to the intervention 4000 jobs were in the backlog, currently there are 250 jobs in the system. Following the intervention the new way of working means that:

- The customer is given a timed appointment at a time that suits them.
- Operatives have the freedom to do the right repair. They have the time to do the job properly, the authority to pull in materials and other support trades as necessary and, importantly, there is due consideration of prevention rather than a short term fix.
- Other repairs are completed during the same visit or an appointment is made.

Value demand has increased from 51 percent to 69 percent with some contractors delivering 89 percent value demand. Customer satisfaction is rated out of 10 and capability charts used to improve failures within the service. The average customer satisfaction score is currently 8.2 with some contractors scoring an average 9 out of 10. More important than the scores is the way in which the feedback is used to understand blockages within the system and remove them. One contractor for example improved satisfaction dramatically by understanding that problems with the appointment system led to disappointed customers. They continued to work on the system until they had wiped out all comments relating to the appointment system. They saw customer satisfaction increase encouraging a move to wipe out the next batch of negative service feedback.

The second intervention involved scrutiny of the entire lettings and allocations process. The work started at the beginning of 2008 and this time involved the use of only three days of consultancy assistance (compared with 30 days on the first intervention). The work has moved the focus away from processing people to helping them. A clear mismatch was identified between supply and demand of properties. In a two-week period the organisation received more requests for housing than they could assist in a year. The result was lots of time spent processing people including tasks such as completing application forms for those that stood no realistic chance of being helped, administering an ever growing waiting list, putting on 'bids' for properties for people that were unlikely to be successful, and answering queries from people wanting to know how they could increase their chances of being assisted.

The redesign seeks to more closely match supply with demand and to provide information on housing options at the point at which an enquiry is received. Applicants for housing receive upfront honest information about housing options and set-up time has been reduced from 30 days to 1 hour. People are now told upfront if they were likely to be helped and time was spent advising on other options as opposed to processing paperwork on them. An emphasis has also been placed on understanding exactly what individuals required and how long this was likely to take.

The other key finding was that many tenancies came at risk of failing in the first 6 months of the tenancy, in the main due to rent arrears. In 2007, nearly half of all tenancies became at risk within the first 6 months. Whilst many tenancies recovered this created much work for the organisation. The redesigned process puts additional time into the value work of setting up a tenancy to succeed. The results of the redesign have demonstrated clear success with tenancies coming under some form of risk in the first 6 months dropping from 43 percent prior to the intervention to a current figure of 14 percent. Alongside this result, the time taken for rent payments to be paid on time has reduced from 31 days prior to the intervention to a current figure of 14 days. The team continues to work to understand the causes of failure and works to reduce the figure bit by bit, understanding and systematically removing the causes of failure.

One year on and some clear results have emerged. Waste work has been cut by 70 percent allowing resources to be focused upon the purpose of the system: 'to offer upfront housing options and to house the right person in the right property at the right time'. Roles have been changed to reduce handoffs and potential for failure and admin systems have been streamlined. Complaints have also fallen dramatically with a move from an average of two complaints a week to a couple per year. The team also feel highly motivated and have the measures to understand what is happening in their system and how it can be improved. Tables 7.2 and 7.3 summarise the key improvements as a result of the two interventions.

Table 7.2   Improvements achieved at charter housing association maintenance and repairs intervention

Charter maintenance and repairs	Before intervention	After intervention
Demand origin	Phone 81 percent In person 7 percent Email/fax/post 9 percent	Not measured
Demand type	Value 51 percent Failure 49 percent *NB Total demand =1776 instances during 5 week check period*	Value 69 percent Failure 31 percent *NB some contractors achieving 89 percent value*
Process steps	Value steps 60 (7 percent) Waste steps 810 (93 percent)	Not measured on ongoing basis (all steps now value)
Measures	No. jobs raised by priority percent completed on target	Permanent (leading*): • percent first time fix • percent appointments given and kept • End-to-end time Permanent (lagging*): • Customer satisfaction Temporary: • percent failure demand
Results		• percent first time fix 100 percent • percent appointments given and kept 87.5 percent Customer satisfaction Averaging 8.2 / 10

* Leading indicates measures needed to know that the organisation is 'on track' to achieving the desired outcome; lagging measures however indicate that the desired outcomes have been attained.

Charter Housing Association currently plans to start work on their third intervention (customer services) in the next few months.

Conclusions

There is evidence of growing interest in systems thinking in the public and third sectors (Chapman, 2004; Seddon, 2008; Advice UK, 2008).

Table 7.3 Improvements achieved at charter housing association allocations and lettings intervention

Charter housing lettings	Before intervention	After intervention
Demand origin	Phone 84 percent In person 12 percent Email 35 percent Post 0.5 percent	Phone 100 percent
Housing applications 1/07-1/08	Application forms 72 percent Web applications 22 percent Phone applications 6 percent	
Demand type	Value 25.9 percent Failure 74.1 percent (System generated 36.1 percent of failure demand)	Waste work cut by 70 percent
Banding (urgent/high/med/low)	Wrongly banded: 50 percent web applications 40 percent application forms	
Set-up time	30 days (720 hours)	1 hour
Rent arrears	• 43 percent of tenants 'at risk' of arrears in first 6 months of tenancy • Time for rent to be paid on time – 31 days	• currently 14 percent 'at risk' • currently 14 days
Complaints	2 per week (104 per year)	2 per year

Housing is one area where some activity has already taken place (OPDM, 2005; NHC, 2006; Jackson et al. 2008; McQuade, 2008). In this chapter we have evaluated secondary evidence of housing organisations in England who have experimented with some success. There is variation in the improvement results, however, and it should be borne in mind that the improvement interventions were 'pushed' onto some of these organisations. Our primary evidence comes from two organisations that 'pulled' a lean systems thinking approach to their organisations. Both organisations have visionary leaders and an inclination towards innovation and experimentation. Constant experimentation is a key ingredient in Toyota's extraordinary success (Spear and Bowen, 1999). It is our contention that the housing sector has an advantage over other parts of the public and third sectors because a good deal of work is already underway. Our aim in this chapter is to draw attention to this work and suggest that it offers a practical way forward for organisations in both housing and the wider public sector looking for improvements in both effectiveness and efficiency. These issues are highly pertinent in the current climate of recession and the demand for public services to achieve more and better service with less funding.

Note

1. It must be noted that in this chapter systems thinking is used as a generic term which refers to the applications of TPS in the service sector.

References

Advice UK. (2008) *It's The System, Stupid, Radically Rethinking Advice*, Report of Advice UK's Radical Advice Project 2007-2008, supported by the Baring Foundation

Bagely, A. and Lewis, E. (2008) *Why Aren't We All Lean?* Public Money and Management, (Chartered Institute of Public Finance and Accountancy), February

Bartezzaghi E. (1999) 'The Evolution of Production Models: Is a New Paradigm Emerging?' *International Journal of Operations and Production Management*, 19(2), 229–250.

Beata, K., Jens, J. D. and Per-Olaf, B. (2007) 'Measuring Lean Initiatives in Health Care Services: Issues and Findings'. *International Journal of Productivity & Performance Management*, 56, 7–24.

Bhasin, S. and Burcher, P. (2006) 'Lean Viewed as a Philosophy'. *Journal of Manufacturing Technology Management*, 17, 56–72.

Bushell, S., Mobley, J. and Shelest, B. (2002) Discovering Lean Thinking at Progressive Healthcare'. *Journal for Quality & Participation*, 25, 20–25.

Chapman, J. (2004) *System Failure: Why Governments Must Learn To Think Differently*. DEMOS: London.

Essex, S., Smith, B. and Williams, P. (2008) *Final Report of the Task and Finish Group*, Commissioned by the Deputy Minister for Housing, WAG.

Flood R. (1999) *Rethinking the Fifth Discipline*, Routledge: London.

Hines, P., Holweg, M. and Rich, N. (2004*)* 'From Strategic Toolkit to Strategic Value Creation: A Review of the Evolution of Contemporary Lean Thinking'. *International Journal of Operations and Production Management*, 24, 994–1011.

Jackson, M., Johnston, N. and Seddon, J. (2008) 'Evaluating Systems Thinking in Housing'.*Journal of the Operational Research Society*, 59, 186–197.

Katayama, H. and Bennett, D. (1996) 'Lean Production in a Changing Competitive World: A Japanese Perspective'. *International Journal of Operations and Production Management*, 16, 8–23.

Krafcik, J. F. (1988) 'Triumph of the Lean Production System'. *Sloan Management Review*, 30, 41–52.

McQuade D. (2008) 'New Development: Leading Lean Action to Transform Housing Services'. *Public Money and Management*, (Chartered Institute of Public Finance and Accountancy), February.

NHC (2006) *A Systematic Approach to Service Improvement - An Update: Evaluation the Sustainability of Systems Thinking in Housing*, Northern Housing Consortium

OPDM (2005) *A Systematic Approach to Service Improvement*, Office of the Deputy Prime Minister

Papadopoulos, T. and Merali, Y. (2008) 'Stakeholder Network Dynamics and Emergent Trajectories of Lean Implementation Projects: A Study in the UK National Health Service'. *Public Money and Management* (Chartered Institute of Public Finance and Accountancy), February

Papadopoulou, T. and Ozbayrak, M. (2005) *Leanness: Experiences from the Journey to Date*, Journal of Manufacturing Technology Management, 16

Radnor, Z. and Boaden, R. (2008) *Lean in Public Services – Panacea or Paradox?* Public Money and Management (Chartered Institute of Public Finance and Accountancy), February

Radnor, Z., Walley, P., Stephens, A. and Bucci, G. (2006) *Evaluation of the Lean Approach to Business Management and Its Use in the Public Sector*, Warwick Business School

Seddon, J. and Brand, C. (2008) *Systems Thinking and Public Sector Performance*, Public Money and Management (Chartered Institute of Public Finance and Accountancy), February

Seddon, J. and Caulkin, S. (2007) *Systems Thinking, Lean Production and Action Learning*, Research and Practice, 4, 9–24

Seddon, J. (2005) *Freedom from Command and Control: a better way to make the work work, the Toyota system for service organisations*, Vanguard Education Ltd.

Seddon, J. (2008) *Systems Thinking in the Public Sector: The Failure of a Reform Regime and a Manifesto for a Better Way*, Triarchy Press

Shingo, S. (1989) *A Study of the Toyota Production System from an Industrial Engineering Viewpoint*, Cambridge, MA, Productivity Press

Spear, S. and Bowen, H. (1999) *Decoding the DNA of the Toyota Production System*, Harvard Business Review, Sep–Oct

Spencer, C. and Guide, V. (1995) *An Exploration of the Components of JIT - Case Study and Survey Results*, International Journal of Operations and Production Management, 15, 72–83

Stake R. (2000) in Denzin and Lincoln (eds.), *Handbook of Qualitative Research*, Sage

Swank, C. (2003) *The Lean Service Machine*, Harvard Business Review, October

Welsh Audit Office (www.wao.gov.uk) February 2006 Inspection of Swansea Housing Association: A Summary Report, Ref 789A2006

Womack J. and Jones D. (1996) *Lean Thinking*, Simon and Schuster, London

Womack J. and Jones D. (2005) *Lean Solutions: How Companies and Customers Can Create Value and Wealth Together*, Simon and Schuster, London

Womack J., Jones D. and Roos D. (1990) *The Machine That Changed The World*, Rawson Associates, NY

Yin R. (2003) *Case Study Research: Design and Methods*, Sage Publications

Zokaei, A. K., Elias, S., O'Donovan, B., Evans, B., Samuel, D., and Goodfellow, J. (2010) *Lean and Systems Thinking in Public Sector in Wales*, Wales Audit Office, Cardiff, UK

8
Systems Thinking for Public Services: Adopting Manufacturing Management Principles

Ayham Jaaron and Chris Backhouse

A management thinking shift has recently been noticed in public services to adopt manufacturing improvement paradigms in their attempt to face economical and operational challenges. This chapter investigates the utilization of systems thinking in public service operations for potential added value. A case study of systems thinking implementation at a UK city council help desk was carried out using in-depth interviews with key personnel coupled with observations and document evaluation. The Organisational Commitment Questionnaire (OCQ) was administered among frontline employees. Results show that systems thinking could create significant added value to the business and to the working place. In addition, a strong relationship was demonstrated between the systems thinking implementation and the affective commitment level of employees. This chapter is one of a few studies that demonstrate the applicability of manufacturing systems in other settings and that they can generate significant added value for the overall service department.

Introduction

Studies of call centre management practices and environment have revealed that there is a tendency to focus on efficiency rather than effectiveness (Dean, 2002), in other words there is a focus on 'quantity' rather than 'quality' (Bain et al. 2002; Mahesh, 2006; Raz and Blank, 2007). This focus on quantity explains such issues as control on employees, high workloads, and less empowerment. Obviously, it is difficult for employees to be customer-oriented in such an environment. This explains one of the reasons for the low levels of service quality in certain call centres (Dean, 2002). It was also found that this focus on quantity at the expense of quality came as a result of a mismatch between what was required from call centres and what was measured within call centres (Robinson and Morley, 2006). To improve service quality in call centres, an integrated approach to service operations construction is required to deliver its anticipated benefits (Schelp and Winter, 2008). However, the creation and implementation of call centres in different organisations have been perceived in many cases as the responsibility of the customer service department without the involvement of other business units. This came in response to the functional specialistion principals applied mainly to large size organisations (Blau, 1971). As a result, employees feel that they are constrained to the unavailability of necessary links with other units when needed (Corea, 2006). Despite this fact, managers in many organisations including public organisations remain committed to 'an ineffective course of action' of continuing executing call centre projects that are failing to achieve a satisfactory level of service quality (McElhinney, 2005). In many occasions, the entrapment to ineffective call centre's projects implementations and management was a result of the application of key performance indicators (KPIs) which are driven by an excessive focus on statistics and regulations to satisfy government targets. This causes the system to hide many repetitive tasks and procedures that are considered as waste. The waste present in the service system makes the service process sluggish and time consuming which in turn harms the customer service level and dissipates monetary resources. However, public service departments have to achieve targets in order to maintain funding (Jaaron and Backhouse, 2009).

Due to the current economic pressures, the expectations of local authorities include a requirement for Value for Money (VFM) for the operations and services they run. This necessitates substantial cashable efficiency savings, ideally without impacting upon service performance, particularly in priority areas. This means that the challenge is to maintain the same service level, if not better, with less resource. All this was built upon in the local authorities' quest for ways to achieve its aim to save money through the application of transformational reviews of systems and managerial regimes.

This chapter explains the gradual shift in management thinking in public sector service departments towards adopting manufacturing support services models as a substitute for traditional functional specialistion. The chapter investigates the impact of the manufacturing support service models on the public services departments and the achievement of twin benefits of significant added value to the business and its employees. In the following sections, the concepts of systems thinking are presented, the concept of *affective commitment* and both its prerequisites and significance on service quality are then discussed. This is followed by a description of the case study organisation and data collection procedures. Finally, results are presented and conclusions discussed.

Lean manufacturing for service operations

Lean manufacturing principles were introduced for the first time by Womack and Jones (1996), based on the Toyota Production System (TPS), as a comprehensive strategy for the elimination of waste from work operations (Ohno, 1988). The elimination of waste is achieved through the creation of value stream maps of operations that deliver products (or solutions in the case of service) (Christopher, 2000; Busi, 2005). Lean manufacturing is defined as the ability of the organisation to do more work with less resources (Christopher, 2000), thus reducing overall costs. It embraces a number of strategies that comprise the whole leanness philosophy, these are JIT (whereby products and parts are only produced when a customer demand is received), workers empowerment, zero inventory, team working, continuous improvement, small production quantities, value streams, and quick systems set-ups (Forrester, 1995).

Nowadays, Lean manufacturing systems are widely used in many manufacturing industries around the globe. They have recently witnessed acceptance from service industries (Robertson, 1999) as a possible strategy to face increasing customer expectations and intensive economic pressures for reduced costs. However, a gap seems to be present in service organisations between the management focus and that of frontline staff. While the premium interest of management in service organisations, similar to other types of organisations, is the cost, the concern used at the customer interface level is of service quality and customer satisfaction (Busi, 2005).

To cover this gap of interest the management has to understand that as the service level increases the operating costs decrease. If the customer receives what he wants from the organisation, then the customer is receiving a quality service with least cost incurred by the organisation as he or she does not need to call again asking for further resources. Likewise, if the organisation is not providing what the customer wants, then most probably the service encounter is poor and the customer is consuming more resources from the organisation since they need to call again until they get what they want. Eventually, if the customer does not get what they want from an organisation, this may cause them to stop using the service and switch over to other competitors (Seddon, 2008). However, the need to satisfy customer demands and reduce the frequency of demand failure requires the elimination of waste in the service systems and the creation of a variety of absorbing operations that can reduce resources consumption and improve capacity. In the context of this chapter, the words 'systems thinking' are used to describe the system that has emerged from the translation of lean manufacturing principles for service departments.

Systems thinking is an approach for the design and management of work. It is based on designing the organisational systems around customer demand instead of in functional hierarchies at which customer demand is analysed over a period of time to collect information about what customers want and expect and what matters most to them. Demand is analysed on the basis of value and failure demand, value demand is what the service department has been established to serve and what the customers want which is of value to them; failure demand is the demand that the service department was not able to serve due to the lack of information or supporting operations. Systems thinking integrates the decision-making processes with the work itself (Seddon, 2005). To design against customer demand is to be more responsive to them. This implies that the waste present in the current system has to be reduced in the new design to enable the quick response. Removing waste implies the redesign of the service processes flow by focusing on minimizing the non-value adding activities from the customer's point of view. When waste is removed the capacity of the system increases which allows for cost reductions and service quality improvements (Seddon, 2003). This way allows for more control because data is in the hands of the people doing the work. Measures used are built in so they automatically tell you what is happening. The result is a self-adapting system. Table 8.1 presents the main features of the systems thinking approach and compares them with the traditional managerial thinking found typically in 'mechanistic' structures (Robey and Sales, 1994).

Service departments are typically exposed to a greater demand variety from the customer than are manufacturing departments (Seddon, 2003). In order for the service organisation to absorb demand variety it needs an adaptive mechanism similar to that of a living organism that can adapt to the surrounding environment in order to function and

Table 8.1 Traditional management thinking vs. systems thinking

Comparative dimension	Traditional management	Systems thinking
Perspective	Top-down	Outside-in
Design	Functional specialisation	Demand value flow
Decision making	Separated from work	Integrated with work
Measures	Budget, activity, targets, output, standards	Related to purpose, variation, and capability
Motivation	Extrinsic	Intrinsic
Management ethic	Manage budget and people	Act on the system
Attitude to customers	Contractual	What matters

Source: Seddon, 2003.

thrive. Such an organic structure is typified by devolved decision making processes (Robey and Sales, 1994). Organisations where employees are given the ability to make work decisions are more able to create a variety absorbing system. In addition people who are working under such standards have a sense of freedom and ownership. The characteristics of this approach are that jobs are wide in scope and employees are allowed to act on a variety of tasks, to learn and to build relationships with customers (Seddon, 2005). These tasks are not governed by rigid rules and procedures; the team shares the responsibility for the work. Hierarchy of control is not usually present thus allowing the team to identify the right person to solve a particular problem. This is congruent to the characteristics of the organic structures introduced by Robey and Sales (1994), and eventually the systems thinking approach is the opposite of 'mechanistic' structures. Help desks, for example, are typically mechanistic structure units by the managerial systems they use. However they are outward-facing entities exposed to the ever-changing, uncertain and demanding surrounding environment, they represent the most intensive and the main channel of interaction with customers (Burgers et al. 2000). Despite the fact that help desks can deal with some demands as value demand, they typically face different demands and conditions than those that are shielded from the environment (e.g., production department, quality assurance department, etc.), the unpredictable demands and conditions increase the uncertainty of the inputs (Robey and Sales, 1994). The emphasis that mechanistic structures must be shielded from the environment strongly indicates that call centres must be given an organic face to improve their value demand dealing capabilities.

Affective commitment (significance and antecedents)

In the past few decades, organisational commitment has emerged as a platform for the identification of the relationship between organisations and their individuals (Commeiras and Fournier, 2001). Mathieu and Zajac (1990) found that the concept of organisational commitment has taken two dimensions in empirical studies, some have described it as a consequence when linking it with work environment, role states, and organisational structures, others describe it as an antecedent when linking it with turnover, absenteeism, and personal behaviour. In the review of literature, organisational commitment has been defined as the employee's psychological attachment to the organisation (Mowday et al. 1979; Meyer and Allen, 1991). Allen and Meyer (1990) argued that organisational commitment which reflects a psychological state has three different components, these are:

Affective commitment

It is the employee's emotional attachment to the organisation. As a consequence, the person strongly identifies with the goals of the organisation and desires to remain a part of it because he wants to do so.

Continuance commitment

The employee commits to an organisation because of the high cost associated with leaving the organisation, including different forms of monetary losses such as pension accruals and social costs such as friendship ties with co-workers. The employee remains a part of the organisation because he has to do so.

Normative commitment

The employee attaches himself to and remains a part of the organisation because of a feeling of obligation. For example, the organisation may have invested resources in training or educating an employee who then feels an obligation to put in a corresponding amount of effort with a view to 'repaying a debt'.

Several studies have demonstrated that organisational commitment, job satisfaction and quality of service are interrelated (Mathieu and Zajac, 1990; De Ruyter, 1999; Malhotra and Mukherjee, 2004).

Among the three components, the affective commitment was found to be more effective than the other two components in influencing the service quality of customer-contact employees. There is also evidence from the relationship marketing literature of the importance of affective commitment in motivating clients and customers to continue their relationship with an organisation. De Ruyter (1999) findings show the availability of different pre-requirements and consequences of commitment in auditor-client relationships. Affective commitment was found to play an important role in this context. He argued that service quality and trust have a positive impact on affective commitment. Customers usually evaluate the overall image of the organisation solely on the basis of their experience with the service encounter and then decide whether to continue with it or not (Brown and Maxwell, 2002).

The impact of affective commitment on service quality can be explained by understanding its antecedents, these were introduced by Mowday et al. (1982) and van Emmrik and Sanders (2005):

- Employee's personal characteristics: if the organisation provided the chance for its employees to fulfil their personal ambitions; desire of achievement, autonomy, and a sense of control on what the employees have, then employees are more likely to develop affective commitment with their employer.
- Organisational structure and job-related characteristics: affective commitment is also related to the employer ability to decentralize decisions making processes to be at the employee's level. This gives employees a feeling of personal importance and value in the organisation. In addition, role clarity and constructive supervisors relationships with employees is particularly important in this regard.
- Psychological contract: The psychological contract refers to the expectations set by employees and employer concerning each other's obligations (Van Emmrik and Sanders, 2005). Employees have expectation of job promotion, employer loyalty and preferences considerations at work. Unmet employee's expectations could result in dissatisfaction and ultimately turnover.
- Work experiences: Employees whose working experiences are rewarding and fulfilled their own aspirations are ready to exert more effort on behalf of the organisation to deliver high levels of service quality than those whose working experiences were less rewarding (Meyer and Allen, 1991; Meyer et al. 1993).

Despite the fact that an ever-changing business environment calls for new forms of organisational structures and management styles, managerial practices and organisational structures commonly implemented in call centres across the service industry can inhibit the development of a rewarding job experience for employees. This is due to standardised work procedures, monitored dialogue (Taylor et al., 2002; Ellis and Taylor, 2006), mechanisation of customer-employee contact, and an emphasis on quantity statistics and targets over the quality of interaction (Mahesh, 2006; Varca, 2006). Eventually employees experience reduced empowerment in making decisions, they also perceive that their values and the ethical norms are not confirming with those of the organisation. As a result employees possess reduced levels of affective commitment. In fact, shared values and ethical norms have been found to be positively related to the development of affective commitment in business relationships (De Ruyter, 1999). However, there are tools to measure the affective commitment of employees; one of these is the Organisational Commitment Questionnaire (OCQ) (Porter et al.1974). A 15 item version OCQ was introduced by Porter et al. (1974) which was further shortened to a nine item version. The shortened nine item version was found to be more superior than the full 15 item version and more effective for measuring affective commitment (Commeiras and Fournier, 2001).

Research questions

A shift has been witnessed recently in management thinking in public services to depart from traditional functional specialisation, this came as a result of the need for new operational models that could potentially achieve the twin benefits of enhanced customer service and reduced overall cost of the business. Therefore, the research questions identified for exploration in this chapter are the following:

RQ1: Are the public service departments are adopting alternative management models applied to manufacturing support services for the management of their operations?

RQ2: Do these alternative models add value to the public service department (both in terms of enhanced customer service and reduced overall operational cost) and help improve the affective commitment of their employees and thus improve their value to their company?

To help answer these research questions, in the following sections, a description of a detailed case study within a UK city council will be presented.

Research site

An independent case study was carried out in the Information and Communication Technologies (ICT) department of Stockport Metropolitan Borough Council in England. The ICT department has a help desk that supports more than 6,000 customers across the council departments and related directorates for their hardware and software needs and IT problem solving. The help desk has a total of 18 employees working on phones and emails. There were two team leaders in the ICT help desk responsible for day to day operations of the work. Employees have varied educational backgrounds, some are college degree holders, some have received IT training and some have taken further education before joining Stockport Council. They are a mixture of young and older employees, most of them are in the age range of 20–35. They are local residents of the area surrounding the council. In general, the work of the employees is very similar to a call centre environment where customers call seeking information on how to solve IT problems or for technical support. The ICT help desk was granted one phone number for all of its customers regardless of the nature of their demand, and with no Interactive Voice Response (IVR) technologies in use. The purpose of the help desk from the customers' perspective was 'to provide customers with IT support and systems they need and will need to do their job effectively'. In contrast, the purpose of the original system from the employees' perspective (derived from the implicit management practices) was 'to do my task and meet the set targets'. This mismatch of perceived purpose was identified by the ICT improvement team as resulting in a sub-optimum solution.

Although Stockport Council was in a good position, significant improvements have been made in recent years: the Council's aspirations for moving from excellent to exceptional require a continuation in the improvement of both performance and the use of resources. Further, the government's expectations include local authorities identifying 3 percent 'cash releasing efficiency gains' each year and demonstrating and embedding VFM in a more explicit manner. The performance of authorities in these areas is measured and reported upon by the Audit Commission, primarily within the use of resources elements of the Comprehensive Area Assessment (CAA). The CAA takes a broader view of VFM and the use of resources, looking not only at the council but at its relationship with partner organisations. Therefore, it was likely that some form of 'step change' will be necessary to meet in full the challenges that lie ahead. In response to the need for improvement initiatives the ICT department was a part of a strategic approach to achieving the improvements and efficiencies necessary to enable the council to demonstrate VFM in the delivery of its priority outcomes, through the transformation of business and service delivery processes and methods. Eventually, the primary objectives of the ICT transformation programme revolved around the generation of substantial cost reductions and efficiency savings, accompanied by performance improvement and increased customer satisfaction, thereby increasing the VFM provided by the council. The transformation programme covered all functions of the ICT department, engaged with all employees within ICT and interacted with a wide cross section of customers who use the services. The transformation programme progressed in three stages:

Check

This stage started with demand analysis. A check team was recruited to perform this crucial stage of the programme, the team was selected on the basis of their ability to deal with demand end-to-end (collectively) and constructively challenge the status quo, in addition to their being respected within and beyond their team. The check team collated information about what customers expect and want from the ICT help desk and what matters to them most. Data collated in this process enabled identification of the major demands coming into the area. A visual representation of each operation carried out in the help desk was developed as a flow chart with three key checks on accuracy being:

- It must be end-to-end (from customer view)
- It must be followed, wherever it goes
- It must capture what the staff actually do

Identification of waste (actions not adding any value from the customer's point of view) present in the service operations flow was then carried out. All processes classified as waste were marked in red on the process flow chart. Processes that add value from a customer's point of view were marked in green.

Redesign

This stage involved defining purpose and new operating principles of the ICT help desk. The team redesigned the processes flow taking what had been learned in the check phase considering the customer 'wants' and then mapping out the service of the future. The new design focused on minimizing non-value adding activities from a customer point of view. However, it was recognised that complete elimination of non-value adding activities from a customer point of view was never going to be possible. A government targets project was set up to meet the reporting expectations of the local authority and inspectorate regimes. The help desk had to report back progress against measures and targets set by the various inspectorate regimes. This project helped the help desk translate targets into permanent measures; measures that relate to the customer purpose and enable the help desk to actively improve the system on an ongoing basis. The new processes were tested, redesigned and re-tested again to make sure that customers get the best possible service before going live in the help desk. The outcome of these new operating principles was productivity improvement in the processing of customers demands; which by implication, resulted in a responsive and positive customer feedback.

Roll-In

This stage covered implementation of the new model within the ICT help desk by a gradual rolling in of employees. As the check team progressed and the discussion was held about the roll-in of staff to this new way of working, it was key to continue the identification of appropriate training. This training included learning about systems thinking and putting that into practice; understanding and using the new ways of working as progressed by the check team. To ensure the ongoing sustainability of the new system design smaller changes to the way of working to improve the service offered was made. This stage involved the identification of further comprehensive staff training needs as they arose in the help desk.

Research methodology

The data was primarily collected through in-depth interviews and questionnaires conducted at the premises of Stockport City Council, followed by observations and documents gathering. Prior to the commencement of interviews a number of emails and a visit have been established with the 'gatekeeper' (Creswell, 2004) to develop a sense of trust as well as to explain the purpose of research. An 'interview protocol' has been prepared as a backup to help in structuring the interviews and taking accurate notes (Creswell, 2004), it consisted of interviewer and interviewee name and position, time and date of the interview, list of questions to be asked and a space where the notes on each question is to be written. A suitable quiet place was arranged by the 'gatekeeper' to conduct the interviews. Sixteen interviews in total were conducted in research site, 11 were frontline employees from the help desk, three middle managers from the ICT department and two senior managers; one holds the services director position and the other is the head of transformation. The 11 frontline employees were interviewed about their working experience before and after the project as a part of a comparison study to explore the changes that happened at the workplace. The remaining interviewees (middle and senior managers) were interviewed about the introduction process of lean systems thinking and the benefits achieved so far at all levels. The purpose of the study and the estimated interview time and how the information of the interview will be treated were all explained to participants before starting the interview. The interviews started with very broad questions about participants roles, responsibilities and general working issues and gradually were narrowed down to more focused issues which are the main concern of the research work, allowing for the employment of the 'funnel interview' (Tashakkori and Teddlie, 1998). To ensure the elimination of the sense of anxiety and discomfort, every interviewee was asked whether they are comfortable with brief note taking and the use of audio tape to record the conversation, with which all interviewees agreed. After completing the interview, interviewees were thanked for their participation and a confirmation for information confidentiality was reassured. Further, the participants were told that a report about the study will be provided for the research site to ensure that results are accessible to every individual concerned. 'Thematic analysis' methods (Taylor and Bogdan, 1984) were employed to identify the main themes constituting the interviewee replies.

The nine item OCQ (Commeiras and Fournier, 2001) was used in the help desk to measure the affective commitment among frontline employees. It used an interval five option likert-type scale with the following anchors ('strongly disagree', 'disagree', 'undecided', 'agree', and 'strongly agree'). An informed consent form was provided as the first part of the questionnaire; a statement which guarantees that the responses from participants will be strictly confidential and data from this research will be reported only in the aggregate. Permission was granted by the head of transformation to start using the questionnaires in the help desk. A web-based questionnaire was sent through the internet to everyone working in the help desk. This provided a quick and easy data gathering and analysis (Creswell, 2004). The frontline employees spent only two minutes on average to complete it, and it did not interfere dramatically with the participants work responsibilities as it was completed in a natural setting. An 'interrupted involvement' (Easterby-Smith et al. 2002) role was adopted during observations; this was done through roaming in the research site over a period of time moving in and out of the help desk, and later documenting the general behaviour atmosphere and the relationship between employees. Documents were also collected through the Head of Transformation and Services Director. These comprised mainly reports about the nature of the project, progress achieved, and the project management plans. They represented a good source of data to make inferences about the management style. They were of particular importance due to their perceived value in discovering things that have taken place before the beginning of the research inquiry (Patton, 2002).

Results and findings

The in-depth interviews conducted were divided into two groups during the analysis stage.

Interview findings related to help desk employees

These are the frontline employee interviews that explore the working conditions under the systems thinking principles, as well as the impact on employees perceptions regarding their working experience. The findings from the thematic analysis (Taylor and Bogdan, 1984) derived the following central themes:

Theme 1: management style

Team leaders and supervisors focus at the help desk shifted from targets and statistics towards percentage of one stop calls and demand analysis, to know what has been done better and how to further reduce repeated phone calls. No phone call recording or monitoring was required and no restrictions were applied on employees to finish a call within a specified time. In addition, no scripts were used for employees to follow when talking to customers. Another primary focus for the management is to make sure that employees are on the phone and that they are ready to help on any phone call if required.

Theme 2: working experience

Employees are now enjoying wider scope of demand which allows for skill development and authority to make decisions on phones. They commented that customers, once they get through, know that they will get what they want in the same phone call. The feedback from customers is very positive. The team share responsibility of the work and informal channels of communication is encouraged to allow for a quicker transfer of knowledge between members. Employees are now getting correct information from customers that could deliver a better service without the need for repeating phone calls.

Theme 3: customer experience and feedback

Interviewees indicted that customers have the facility of reporting their opinions and views about the help desk service through phones and emails. Customer feedback has been very positive. Customers get what they want with the elimination of transfer from pillar to post. One employee is now dealing with the demand in a very efficient way with more time to speak to customers. No IVR technology is used in the help desk with the availability of only one calling number. Employees indicated that 85 percent of incoming phone calls are now dealt with 'one stop'.

Theme 4: performance measurement

Employees are measured and evaluated on the basis of sticking to working principles of meeting customer demand. Team leaders log into the systems to track each employee profile on a daily basis to see the frequency of phone calls that have been met one stop, each employee is expected to handle at least five calls every day and complete them 'one stop'. Employees commented that it is now possible to complete phone calls 'one stop' due to the authority they have to make decisions and deliver the optimal solution required. In addition, the correct information collected from customers allowed for the precise identification of problems and thus a satisfactory solution to be provided.

Theme 5: systems thinking contributions to workplace

Interviewees regarded the following as the most important contribution to the help desk after the implementation of systems thinking:

- Identification of customer problems very quickly
- Clarity on the system due to the continuous demand analysis
- One stop handling of one or more customer demands
- Focus on what customer wants and not caught by procedures and targets
- System waste (unproductive processes) elimination and performance transparency
- Freedom to act on the system

Theme 6: barriers against implementing systems thinking in other organisations

Most of the employees claimed that lack of knowledge about systems thinking and the lack of investigating their current systems performance were behind the limited utilization of these systems in other public service departments and organisations. They indicated that if managers study the flow of processes in their systems they will find out that the systems are hiding a lot of waste and operational problems, and that probably could be the main endeavour to convince them of systems thinking instead of traditional functional specialisation. Other reasons indicated by employees were the ignorance of the customer when designing service systems and the managers need to protect a bad system because it is their own achievement. Other reasons indicated by employees were the fear of change and losing control on the systems, as this changes the whole working philosophy. However, some very skilled employees commented that they opposed the new systems at the beginning because it added trivial tasks to their daily work, this was considered, in their point of view, a source of deskilling practise and not a job enrichment.

Interview findings related to senior and middle managers

The senior and middle managers interviews explored the introduction process of the new systems to the ICT department and significant potential for value creation. A thematic analysis for these interviews generated the following themes:

Theme 1: departmental integration

Managers recognised communication between departments at the city council as important. However, they indicated that the functional specialisation model created a silo between sections and departments as every department was viewed as a separate entity that should not interfere with the work of other departments, one manager stated, 'we thought functional specialistion was an efficient way to do the work but it was not...the system was frustrating and did not allow for an open door policy'. All managers claimed that they have witnessed huge improvements in communication, formal and informal meetings at the manager's level. This allowed for significant information sharing that was necessary to streamline service operations for better customer service. Also meetings with other teams and directors in other departments were carried out to share the effects of the new system on them in a relevant way.

Theme 2: how to justify systems thinking as a counterintuitive initiative

All managers regarded the following contributions of systems thinking as important tools to justify it as a counterintuitive initiative in public service departments:

- Cutting down the waste in the service system that makes the service process sluggish and time consuming
- Saving resources and money without cut in service
- Passing on calls from pillar to post was too high in the old system. Currently, 85 percent of calls are done 'one stop' by one employee for each call.
- Systems thinking is the only model that works with the 'human side' to change the nature of work
- Customer feedback and percentage of calls done first time
- Reduction in repeated calls and thus improved productivity
- Systems clarity and transparency
- Focus on the job rather than maintaining the system

Theme 3: communication plan for implementation

Four managers indicted that they had a communication plan to use with their peers and subordinates that accompanied the project. The communication with the subordinates was done on a daily basis and was informal most of the time, whilst communication with peers and higher level management was formal and was done on a regular basis, a senior manager commented, 'I learned to tailor the message according to the type of receiver...for a person who is 'factual' I tell him that ROI was in 9 months...for other types of people I invite them to come and have a look'. Mangers communicated with government funding bodies, council's senior board level, council's elected members, other departmental managers and trade unions for consultation.

Theme 4: tools and strategies to get management approval

Knowledge about systems thinking and awareness about its benefits to the business were discussed. Managers found that knowledge and awareness are key steps in this process, they indicated that top management involvement in a fundamentals course about systems thinking is very vital to change the way they think about their current systems. They also indicated that quantitative measures of the new working model would play a major role in the process of securing top management support. Quantitative measures should indicate improvements in efficiency and demonstrate VFM, for example they used measures like:

- 85 percent of demand now is met and dealt with 'one stop' against 17 percent demand that was met before.
- It takes less than a day now to fix a call against 11 days for the same type of call.
- Fewer people are required to do more jobs, this is an opportunity to remove some agency workers and save money.

Theme 5: capacity building

All managers indicated that as a part of the development of 'expert' internal capacity, an educational element for interventionists was employed. This was called core curriculum, it gave underpinning knowledge to colleagues who are being developed to be potential lead interventionists in the future. The first programme has been completed with the intention of running future educational elements to give this cohort the opportunity to develop their competence as lead interventionists. To build curiosity among other departmental managers around 100 internal managers were invited to take the fundamental educational element course on systems thinking to spread the word. This was followed by some follow-up visits to their participants departments to capture any opportunity in the organisation to make a systems intervention elsewhere.

Findings regarding affective commitment

The affective commitment of employees was measured using the nine item OCQ. Responses are sought from statements such as 'I talk up this council to my friends as a great organisation to work for' and 'this council really inspires the best in me in the way of job performance'. A total of 18 employees working at the help desk were available at the time of the questionnaire, all employees responded targeting a 100 percent response rate. The data collected from this questionnaire was analysed to examine how affectively committed the help desk employees were. Responses to the nine items are averaged to obtain a single score for each item; the standard deviation for each item was also calculated (Table 8.2). An overall mean for the nine items of 3.77 was achieved. According to Porter, et al. (1974) a return below 3.0 would indicate employees' tendency to leave their organisation as a result of low affective commitment levels, while a return of 3.0 would indicate a neutral level of affective commitment and that values of 3.5 are typical in many organisations. This provided a clear indication of a fairly high affective commitment level among employees in the help desk. This value shows

Table 8.2 Mean and standard deviation for affective commitment questionnaire

Item	No.	Minimum	Maximum	Mean	S.D.
Q1: I am willing to put in a great deal of effort beyond that normally expected for this council to be successful.	18	3.00	5.00	4.0588	0.8726
Q2: I talk up Stockport Council to my friends as a great organisation to work for.	18	2.00	5.00	3.8125	1.0740
Q3: I would accept almost any type of job assignment in order to keep working for Stockport Council.	18	1.00	4.00	2.9375	1.0289
Q4: I find that my values and Stockport Council's values are very similar.	18	2.00	5.00	3.8125	1.0178
Q5: I am proud to tell others that I am part of this council.	18	2.00	5.00	3.8750	1.1143
Q6: This council really inspires the best in me in the way of job performance.	18	1.00	4.00	3.6250	1.0416
Q7: I am extremely glad I chose Stockport Council to work for over others I was considering at the time I joined.	18	2.00	5.00	4.0000	1.0226
Q8: I really care about the fate of this council.	18	3.00	5.00	4.2500	0.7859
Q9: For me, this is the best of all councils for which to work.	18	2.00	5.00	3.5625	1.0431
Overall mean				**3.77**	
Internal consistency (coefficient α)				**0.94**	

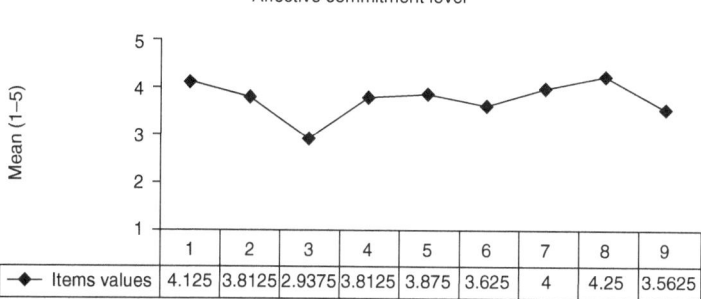

Figure 8.1 Affective commitment of help desk staff

that on average all the respondents agreed on the questionnaire items. A summary of the results is shown in Figure 8.1.

Analysis and discussion

Value for employees

The results reported above show a dramatic change in the philosophy of work as compared to the traditional office values found in the majority of service departments in the government sector. The management style usually found in service departments, including help desks, is focused on achieving job targets and minimising costs, the internal work is designed around the 'Principles of Taylorism' (Bain et al. 2002; Batt and Moyniham, 2002; Cartwright, 2003). This approach is based on standardising work procedures in which the employees need to handle demand in a repetitive manner with detailed descriptions of procedures and work standards. The standardisation of procedures is perceived to increase the mechanisation of the customer-employee contact and decrease service flexibility. Help desk employees following this kind of model will lose a large measure of control over their self-presentation to customers. This will leave them with little autonomy in negotiating their interactions with customers (Deery et al. 2002). Arguably, the traditional help desk environment represents the latest form of the Taylorist principle as it is a common trend in help desks globally to practice high levels of monitoring their employees (Armistead et al. 2002).

However, the results of interviews conducted with frontline employees at the help desk have shown that the implementation of the systems thinking approach has delivered numerous improvements. Employees under the new system are no longer restricted to repetitive job handling procedures or target achieving dilemmas, they are empowered to do the job in the best way they see is vital to satisfy customer needs. Hence, employees have opportunities to develop their working skills by handling a wide range of challenging demands on a daily basis. Their performance is evaluated on their ability to help the customer solve his problem from the first interaction without the need for the customer to call again. Further, employees operate as a team that shares the work responsibility; an employee can seek support from a more skilled colleague to solve a customer problem while on the phone. Obviously, employees working in this environment have a feeling of belonging and ownership of the workplace; they have the freedom to make decisions to provide high quality service in a relatively short time, and they enjoy the open channels of communication between themselves as well as other departments. Employees, thus, have a sense of freedom to act and excel in their jobs.

When linking these job characteristics and environment offerings for employees with the antecedents of affective commitment discussed earlier, an expectation of high affective commitment level among employees can be concluded. The nine item OCQ provided a value of 3.77 for the affective commitment level among employees that reflected the high level expected (Porter et al. 1974). This indeed proved the value-added to frontline employees' life in the help desk in terms of work experience and job-related characteristics.

Value for managers

Help desks represent an organisational structure with relatively few layers of management where managers comprise 12 percent of the employees (Holman et al. 2007). They are top-down flat organisations (T-form organisations) operating in a functional specialised parent organisation (Holman et al. 2007; Adria and Chowdhury, 1999) with a rigid management model that controls employees behaviour to match the requirements of management, monitors their activities and intensively focus on achieving work targets and reduce operational costs (Seddon, 2008). However, these management practices do not provide clarity on system operations, they also cannot indicate whether the demand received was resolved by providing a reasonable solution that satisfies a customer.

The results achieved from interviewing middle and senior managers in the ICT department indicate that systems thinking has provided clarity on the system due to the continuous demand analysis which helped managers identify potential problems in the services offered and thus the immediate corrective measures to be taken. In addition, system clarity and continuous improvement have allowed managers to focus on the main purpose of the help desk (i.e. supporting customers) rather than maintaining the system against failures. Systems thinking was found to eliminate the waste in operations that helped managers achieve automatic productivity and capacity improvements, customers do not need to call again which allowed employees to handle more demands in an efficient way without a cut in service, and eventually reduce resources consumption and overall costs. Significantly, the employment of an economical way to improve the work was another contribution given to managers by systems thinking, managers indicated that systems thinking deals with the human side to change the nature of work without the need for any tools or technologies. Further, the creation of a committed frontline of employees in the help desk has significantly resulted in reducing the burden of managing people's behaviour in the help desk as the general moral system of the workplace controls the human resources behaviour and not the traditional top-down hierarchy.

Value for customers

Many organisations have realised that the relationship with customers should not end at the moment they receive their demand (Feinberg et al. 2000). It is believed that organisations should provide communication channels that can add value to the customers' experience with the organisation (Marisco, 1996). As a result, call centres have emerged as an important port for customer relationship management activities (Kotorov, 2002) and the call centre employees are seen as the main link between organisations and customers (Burgers et al. 2000). However, although the 'caller behaviour is difficult to objectively measure' (Betts et al. 2000), they evaluate the quality of the service on the basis of the service encounter itself and the expected benefits they can achieve from that encounter. Therefore, it is important for organisations to pay attention to the structures and quality of their services to achieve the value added and satisfaction in their customer interactions. The results of interviews at both levels (i.e. frontline employees and managers) reinforced the strong association of systems thinking and customer satisfaction. This came as a result of the rapport initiated between customers and employees, customers are not transferred from pillar to post anymore, only one employee handles the customer demand, they are now enjoying a solution delivery service in less than a day as compared to

11 days before the systems thinking. Apparently, the affective commitment of employees accompanied with system waste elimination created a difference for the customer experience and added value to the service encounter; no unproductive processes are used anymore and employees are willing to exert more effort on behalf of their organisation to deliver excellent service.

Value for business

One may argue that managers may know from targets and statistics whether people are working according to requirements or not. However, employees may rush to achieve their target for answering their customers' demands with low abandonment rates, but the question remains: are they doing the right work whilst having customers on the phone? Are they communicating in the manner that reflects the company's values and are they achieving the contact benefits? Employees may learn how to cheat their numbers to avoid management attention (Seddon, 2008), they can provide fast service even though the customers may be misunderstood and their information incorrectly entered (Cleveland, 2006). Thus, business resources can be lost due to poor service delivery. Interviews with managers have identified that systems thinking has provided a workplace with a relaxed environment where service encounters can be achieved without the need to stick to a pre-specified time allowance. This inevitably helped improve solution delivery processes, prevented calls repetition, and eventually allowed for resources savings. Another significant dimension of interest is the departmental integration, open channels of communication between the ICT help desk and other departments affected by its work were established. Formal and informal communication at the management level allowed for significant information sharing. Interestingly, this created consistency in the services offered by other departments through streamlining service operations. Further, systems thinking made it possible for management to identify the opportunities for making cost savings and performance improvements in the short and medium term, both from a corporate and service perspective. The waste elimination element of the system was viewed as a resources saving activity that used to be a major cause for capacity reduction. In addition, systems thinking allowed the ICT help desk to do more with less, fewer people are required to do more jobs, creating an opportunity to remove some agency workers and save money.

Implementation roadmap: Initial model

Interviews with senior and middle managers allowed the presentation of all the activities carried out to implement the project in the ICT help desk. An implementation roadmap that may be used by managers in other public service departments was generated as a result (Figure 8.2).

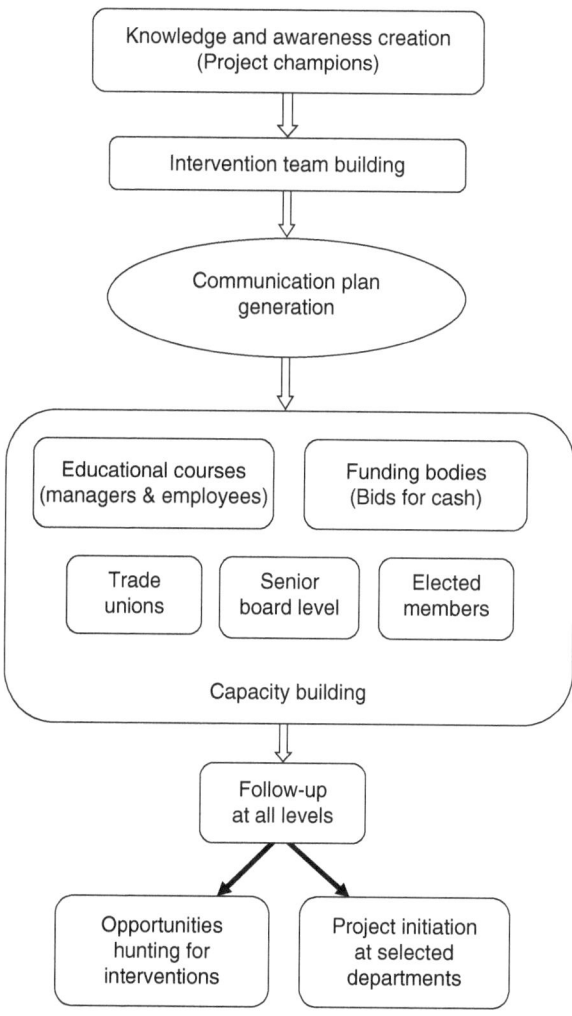

Figure 8.2   Systems thinking implementation roadmap: initial model

A small-scale pilot study of the initial roadmap was shared with senior and middle managers, who were the core intervention team members at the research site. .

Managers have emphasised that the project was of a highly sensitive nature as communication was required at different levels throughout the project. The communication plan included justification of a systems thinking project and the problems discovered in the current system due to the initial demand analysis. As a convincing tool, quantitative measures such as service delivery times, percentage of one stop calls, and customer surveys were also used in the communication process. Government funding bodies, senior board level, elected members, trade unions, and other department managers were targeted in this process. A lack of knowledge about systems thinking and the hidden problems of the current system, in addition to the ignorance of customer voice when designing service systems were identified by interviewees as the main barriers against systems thinking implementation. Therefore, interviewees have indicated that knowledge and creating awareness about systems thinking among managers and other bodies in charge were of great importance, educational presentations and courses were organized and top managers and peers were invited, this has helped build curiosity among people and thus capacity building in the council to support the project.

Conclusion

The fundamental objective of this research has been to investigate the utilization of lean manufacturing principles in public sector services. In this context, the words 'systems thinking' were employed to describe lean manufacturing principles for service departments. It is based on designing service operation systems around customer demand instead of in functional hierarchies.

The analysis of the results achieved in this case study clearly answers the research questions posed in this chapter. With respect to the first research question, it has been indicated that managers in public services are realizing the need to employ more innovative interventions to achieve financial savings and better performance. As a result a shift has been noticed recently in management thinking in public services to adopt systems engineering models utilized in the manufacturing sector, this has occurred due to the economical and governmental pressures exerted on councils and public services to generate substantial cost reductions and efficiency savings, accompanied by performance improvement and increased customer satisfaction. With respect to the second research question, the evidence from this research indicates that the utilization of the systems thinking approach to design services in the public departments has significantly contributed to the achievement of added value to employees, managers, customers, and eventually business. A clear strategy has therefore emerged between the use of the systems thinking approach in call centres and the contributions to the public departments that embrace it. Continuous demand analysis provided managers with clarity on the system that allowed managers to focus on supporting customers rather than maintaining the system against failures. It has also become clear that the use of systems thinking to design call centre operations is likely to increase the affective commitment of frontline employees and thus decrease turnover rates and absenteeism. Higher levels of employees' affective commitment created a difference for the customer experience; employees are willing to exert more effort on behalf of their organisation to deliver excellent customer service. This suggests that the ability of employees to control the work and to decide on the way they handle and receive information is a key factor of improving business productivity and service quality.

It cannot be considered easy to establish lean manufacturing principles (i.e. systems thinking) in public service departments, as the methods and the benefits are often counterintuitive. The challenge against the establishment of such a counterintuitive initiative, within the highly controlled environment of government departments, is to convince decision-makers that this is the way forward for their organisation. This research takes some initial steps towards the creation of a roadmap that could secure top management support.

References

Adria, M. and Chowdhury, S. (1999) 'An Organisational Design Approach of Implementing Call Centres and Intranets for Improving Customer Service'. *Administrative Sciences Association of Canada, Annual Conference,* 20(4), 22–33.

Allen, N. J. and Meyer, J. P. (1990) 'The Measurement and Antecedents of Affective, Continuance and Normative Commitment to the Organisation'. *Journal of Occupational Psychology; Leicester,* 63(1), 1–18.

Armistead, C., Kiely, J., Hole, L. and Prescott, J. (2002) 'An Exploration of Managerial Issues in Call Centres'. *Managing Service Quality,* 12(4), 246–256.

Bain, P., Watson, A., Mulvey, G., Taylor, P. and Gall, G. (2002) 'Taylorism, Targets and the Pursuit of Quantity and Quality by Call Centre Management'. *New Technology, Work and Employment,* 17(3), 170–185.

Batt, R. and Moyniham, L. (2002) 'The Viability of Alternative Call Centre Production Models'. *Human Resources Management Journal,* 12(4), 14–34.

Betts, A., Meadows, M. and Walley, P. (2000) 'Call Centre Capacity Management'. *International Journal of Service Industry Management*, 11(2), 185–196.

Blau, P. M. (1971) *The Structure of Organizations*. Basic Books: Ney York.

Brown, G. and Maxwell, G. (2002) 'Customer Service in UK Call Centres: Organisational Perspectives and Employee Perceptions'. *Journal of Retailing and Consumer Services*, 9(6), 309–316.

Burgers, A., Ruyter, K., Keen, C. and Streukens, S. (2000) 'Customer Expectation Dimensions of Voice-to-voice Service Encounters: A Scale-development Study'. *International Journal of Serviced Industry Management*, 11(2), 142–161.

Busi, M. (2005) *'Seeing the Value in Customer Service'*, Working Paper, The Centre for Business Process Outsourcing, University of Strathclyde, Glasgow, Scotland, UK.

Cartwright, S. (2003) 'New Forms of Work Organisation: Issues and Challenges'. *Leadership and Organisation Development Journal*,24(3), 121–122.

Christopher, M. (2000) 'Supply Chain Migration from Lean and Functional to Agile and Customised'. *Supply Chain Management: An International Journal*, 5(4), 206–213.

Cleveland, B. (2006) *Call Center Management on Fast Forward: Succeeding in Today's Dynamic Customer Contact Environment*. ICMI Press: Annapolis.

Commeiras, N. and Fournier, C. (2001) 'Critical Evaluation of Porter *et al.*'s Organisational Commitment Questionnaire: Implications for Researchers'. *The Journal of Personal Selling and Sales Management*, 21(3), 239–245.

Corea, S. (2006) 'Mounting Effective IT Based Customer Service Operations under Emergent Conditions: Deconstructing Myth as a Basis of Understanding'. *Information and Organization*, 16(2), 109–142.

Creswell, J. W. (2004) *Educational Research: Planning, Conducting, and Evaluating Quantitative and Qualitative Research*, Second Edition, Pearson Education: Ohio.

De Ruyter, K. (1999) 'Commitment in Auditor–client Relationships: Antecedents and Consequences'. *Accounting, Organizations and Society*, 24(1), 57–75.

Dean, A. M. (2002) 'Service Quality in Call Centres: Implications for Customer Loyalty', *Managing Service Quality*, 12(6), 414–423.

Deery, S., Iverson, R. and Walsh, J. (2002) 'Work Relationships in Telephone Call Centres: Understanding Emotional Exhaustion and Employee Withdrawal'. *Journal of Management Studies*, Vol. 39, No. 4, pp. 471–496.

Easterby-Smith, M., Thorpe, R. and Lowe, A. (2002) *Management Research: An Introduction*, Second Edition, London: Sage Publications.

Ellis, V. and Taylor, P. (2006) 'You don't know what you've got till it's gone: re-contextualising the origins, development and impact of the call centre', *New Technology, Work and Employment*, Vol. 21, No. 2, pp. 107–122.

Feinberg, R., Kim, I., Hokama, L., De Ruyter, K. and Keen, C. (2000) 'Operational determinants of caller satisfaction in the call centre', *International Journal of Service Industry Management*, Vol. 11, No. 2, pp. 131–141.

Forrester, R. (1995) 'Implications of lean manufacturing for human resource strategy', *Work Study*, vol. 44, no. 3, pp. 20–24.

Holman, D., Batt, R. and Holtgrewe, U. (2007) *'The global call centre report: International perspectives on management and employment'*, Global Call Centre Research Network, United Kingdom.

Jaaron, A. and Backhouse, C. (2009) 'Affective commitment stimulation through systems thinking'. Paper presented at the *QUIS 11 International Conference. Moving Forward With Service Quality*. June 11–14, 2009. Wolfsburg, Germany.

Kotorov, R. P. (2002) 'Ubiquitous organisation: organisational design for e-CRM', *Business Process Management Journal*, Vol. 8, No. 3, pp. 218–232.

Mahesh, V. S. (2006) 'Improving call centre agent performance: A UK-India study based on the agent' point of view', *International Journal of Service Industry Management*, Vol. 17, No. 2, pp. 136–157.

Malhotra, N. and Mukherjee, A. (2004) 'The relative influence of organisational commitment and job satisfaction on service quality of customer-contact employees in banking call centres', *Journal of Services Marketing*, Vol. 18, No. 3, pp. 162–174.

Marisco, K. (1996) 'Call Centers: Today's New Profit Centers '. *AT&T Technology*, 10(4), . 14–18.

Mathieu, J. E. and Zajac, D. M. (1990) 'A Review and Meta-analysis of the Antecedents, Correlates, and Consequences of Organisational Commitment'. *Psychological Bulletin; Washington*, 108(2), 171–194.

McElhinney, D. (2005) 'Concept of Entrapment and Decision-making'. *Management Decision*, 43(2), 189–202.

Meyer , J. P. and Allen, N. J. (1991) 'A Three-component Conceptualization of Organisational Commitment'. *Human Resources Management Review*, 1(1), 61–89.

Meyer, J. P., Allen, N. J. and Smith, C. A. (1993) 'Commitment to Organisations and Occupations: Extension and Test of a Three-component Conceptualization'. *Journal of Applied Psychology*, 78(4), 538–551.

Mowday, R. T., Porter, L. W. and Steers, R. M. (1982) *Employee-Organisation Linkages: The Psychology of Commitment, Absenteeism, and Turnover*. Academic Press: New York.

Mowday, R. T., Steers, R. M. and Porter, L. W. (1979) 'The Measurement of Organisational Commitment'. *Journal of Vocational Behavior*, 14(2), 224–247.

Ohno, T. (1988) *Toyota Production System: Beyond Large-Scale Production*, Productivity Press: New York.

Patton, M. Q. (2002) *Qualitative Research and Evaluation Methods*, Third Edition, Sage Publications: London.

Porter, L. W., Steers, R. M. and Boulian, P. V. (1974) 'Organizational Commitment, Job-Satisfaction, and Turnover among Psychiatric Technicians'. *Journal of Applied Psychology; Washington*, 59(5), 603–609.

Raz, A. and Blank, E. (2007) 'Ambiguous Professionalism: Managing Efficiency and Service Quality in an Israeli Call Centre'. *New Technology, Work and Employment*, 22(1), 83–96.

Robertson, M. (1999) 'Application of Lean Production and Agile Manufacturing Concepts in a Telecommunications Environment'. *International Journal of Agile Management Systems*, 1(1), 14–17.

Robey, D. and Sales, C. A. (1994) *Designing Organizations*, Fourth Edition, McGraw-Hill/Irwin: USA.

Robinson, G. and Morley, C. (2006) 'Call Centre Management: Responsibilities and Performance'. *International Journal of Service Industry Management*, 17(3), 284–300.

Schelp, J. and Winter, R. (2008) 'Business Application Design and Enterprise Service Design: A Comparison'. *International Journal of Services Sciences*, 1(3/4), 206–224.

Seddon, J. (2008) *Systems Thinking in the Public Sector*, First Edition, Triarchy Press: Axminster, UK.

Seddon, J. (2003) *Freedom from Command and Control: A Better Way to Make the Work Work, Buckingham*. Vanguard Education Ltd: *England*.

Seddon, J. (2005) 'Freedom from Command and Control', *Management Services; Enfield*, 49(2), 22–24.

Tashakkori, A. and Teddlie, C. (1998) *Mixed Methodology :Combining Qualitative and Quantitative Approaches*, Sage Publications: *London*.

Taylor, S.J. and Bogdan, R. (1984) *Introduction to Qualitative Research Methods: The Search for Meanings*, John Wiley & Sons: New York

Taylor, P., Mulvey, G., Hyman, J. and Bain, P. (2002) 'Work Organization, Control and the Experience of Work in Call Centres'. *Work, Employment and Society*, 16(1), 133–150.

Van Emmrik, I. J. Hetty and Sanders, K. (2005) 'Mismatch in Working Hours and Affective Commitment: Differential Relationships for Distinct Employee Groups'. *Journal of Managerial Psychology*, 20(8), 712–726.

Varca, P. (2006) 'Telephone Surveillance in Call Centres: Prescriptions for Reducing Strain'. *Managing Service Quality*, 16(3), 290–305.

Part II

Systems Thinking in Private Sector

9

Real Service Improvement: An Empirical Investigation of Service Improvement Initiatives within a UK Bank

David Longbottom, Julie Hilton and Ying Xia-Zheng

This chapter investigates the outcomes of service improvement initiatives undertaken within a major UK bank. It reports on the processes and outcomes from the perspective of bank employees engaged with the change initiatives. The chapter draws on the data to make recommendations for a shift in focus and a new approach Primary research consists of a survey of bank employees (sample 224) followed by depth interviews and observations. The research finds that bank change initiatives have focused on particular change models in recent years, and that there is a general feeling that these, in the majority of cases, have not achieved the expected results. The chapter investigates the reasons for this and draws on the evidence to suggest alternative approaches which are more grounded in empirical studies.

Introduction and Context

The banking sector is going through a severe period of change and restructuring. This chapter concentrates on service improvement initiatives in a major UK bank, from 2006 to 2008. At the commencement of the study the financial environment was stable and prosperous; but by 2008 the events of the 'credit crunch' had started to be apparent.

The majority of service improvement and change initiatives were derived from the use of three particular change models:

- European Business Excellence Model (EBEM), The European Foundation for Quality Management, 1990 onwards

- Balanced Scorecard (BS), Kaplan and Norton, 1996 onwards
- SERVQUAL, Parasuraman, Zeithaml, and Berry, 1985 onwards

Preliminary findings showed that of 64 change initiatives identified over the period, 55 involved the use of EBEM/BS (quality based solutions), and 9 involved SERVQUAL (marketing based solutions). Discussions with employees revealed a perception that quality and operational projects were not well aligned with marketing activity and this became a central theme for our study; to understand why this might be, and the consequences for service improvements.

This chapter reports on the process, experiences, and outcomes from the perspective of employees engaged in the initiatives.

Literature summary

The review of literature presents a summary and analysis of key factors in service improvement, from the perspective of two disciplines, which we describe very broadly as quality based solutions, and marketing based solutions. This provides a theoretical and conceptual basis for the research design and subsequent analysis by identifying critical factors which may be evident within successful service improvement initiatives. Two major themes are identified; integration of effort for service improvement; value adding and the nature of services.

Integration of effort for service improvement

Oakland (2001) has long argued that TQM starts with marketing. Sila and Ebrahimpour (2003) find 76 empirical studies on critical factors in TQM, and show that customer focus is identified in 53, ranking second only behind leadership and commitment (67). Similar positions have been reached in empirical studies by Saraph et al. (1989), Bossink et al. (1992), Porter and Parker (1993), and Black and Porter (1996).

Within the marketing literature (Piercy, 2008) presents that the role of modern marketing can no longer be confined to the activities of the marketing department, and finds that marketing to employees is a key task in strategic planning which should not be taken for granted. Similar positions are reached by Doyle (2001), and MacDonald (2003). Ahmed and Rafiq (2003) argue that organisations need to better employ marketing techniques internally to staff. Similar arguments can be found within the internal marketing and services literature from Gronroos (1981); Berry and Parasuraman (1991); Sargent and Saadi (1998); Varey and Lewis (1999); and Ballantyne (2003).

Morgan and Piercy (1998), however, concluded that marketing did not play an important role in quality. They suggest that functional organisation structures, and narrowly defined roles, may have contributed to this. Some authors have developed alternative organisation structures; for example Kotler (2003) sees the organisation as a series of rings with marketing at the centre; Piercy (2008) proposes a value adding structure with a central going to market process; Macdonald (2003) presents a matrix style organisation. We find, however, very little empirical work to support such models and consequently little evidence of practice.

Mele (2007) reaches a similar position, identifying that whilst TQM and marketing are complementary and synergistic in enabling value creation, there is 'a paucity in the literature on the relationship between the two domains in value creation'.

Some insights into practical and implementation issues are presented by Seddon and Caulkin (2007). They propose that implementing a systems thinking approach starts with understanding customer requirements and they go on to describe a methodology for identifying how this might be achieved, developed, and implemented. Similar systems thinking examples can be found in Checkland (1997); Ackoff (1999); and Jackson (2003).

The routes of such fundamental ideas can also be traced back over many years and are inherent for example in Ohno (1978), presenting the Toyota Production System (TPS) that is based on customer demand and flow; and Womack et al. (1990) in 'The Machine That Changed the World'. Some common characteristics emerge, for example; an emphasis on understanding value from the customer perspective, a focus on the work, aligning roles to manage variety in demand, and the delivery of value. Organisation structure appears to be a consequence of change rather than a fixed component. It might be inferred from this approach to systems that integration of effort is more likely to be achieved where systems are looked at in their entirety (or end to end), rather than within functional silos or from the viewpoint of separate disciplines.

Seddon and Caulkin (2007) warn that systems thinking ideas predominantly developed within manufacturing environments may not easily transfer to services environments, and are critical of some practices occurring within the UK public sector.

Value adding and the nature of services

The notion of value and value adding has been prominent within marketing literature over the past decade. In summary we find the emerging key factors include: managing internal relationships; reciprocal relationships; building understanding and intimacy; trust and commitment; (Ahmed and Rafiq, 2003) knowledge renewal based on mutual exchanges; learning activity; and market relevance; (Ballantyne, 2003) a process by which value is profitably created for internal and external customers (Varey and Lewis, 1999; deChernatony et al. 2000; Doyle, 2001; Davis, 2001; Piercy, 2008; MacDonald, 2003). Three critical components are commonly identified by Webster (1992); deChernatony et al. (2000); Piercy (2009); and Mele (2007) as value definition, value development, and value delivery.

Anderson et al. (2006) argue that there is little evidence of methodology for identifying value propositions that resonate with customers. They find that managers often construct value propositions by simply listing product and service benefits (all benefits approach) or by identifying strengths against competitors (favourable points of difference). Both methods they contend are flawed and fail to identify value from the customer's perspective, the points of difference that are important to each individual customer (or resonating focus as they call it). This idea of resonating focus has particular relevance we would submit within services industries, where customers have different priorities, and as Seddon and Caulkin (2007) have pointed out, systems need therefore to be able to handle variety in demand, and locate skills at the point of service delivery. This notion however is often counterintuitive to the way managers have come to think about organisation structures and systems, arranging work into silos with an emphasis on repetitive specialised operations requiring less skill (and thus implicitly requires a fundamental shift in mindsets and behaviour).

Service quality has been central to service marketing and management literature with strong links made to customer satisfaction, loyalty and profitability, Parasuraman et al. (1985), Heskett et al. (1997). The distinctive nature of the service quality literature evolves from the general recognition of the nature of service (heterogeneity, intangibility, perishability, and simultaneity) and significance of the interactive nature of the service offering. The importance of the interplay between customer and organisation has given the definitions of service quality a customer focus. Quality is 'whatever the customer perceives it to be' (Gronroos, 2007 p. 73) moving the service perspective on quality away from the traditional definitions within quality literature. The research has focused on understanding the process the customer goes through in interacting with the organisation. The technical and functional model proposed by Gronroos (1984) identified two aspects of quality but gave particular importance to the functional dimension (the how of being served) because 'an excellent service process creates a distinct and sustainable competitive edge' (Gronroos, 2007 p. 71).

Much of the literature concentrates on how the customer derives a perception of service quality through the service encounters with the organisation. Indeed Normann (1992) considered there were moments of truth when the consumer perception of service quality might change.

The conceptualisation of customer value is grounded in the widely accepted disconfirmation theory which posits that perceived quality is a function of customer expectation of the service measured against perception of service performance. Whilst not without criticism (Carrillat et al. 2007 offer a meta analytical view) the dominant and widely used SERVQUAL (Parasuraman et al. 1985) suggests attributes (the quality dimensions) the customer looks for in assessing service quality and has been widely used by large organisations to measure perceived service quality. The quality gap model (Parasuraman et al. 1985) is designed to offer insights into the possible causes of a gap between customer expectation and customer perception of quality.

In a meta analysis Seth et al. (2005) review 19 service quality models that are intended to enable management to identify quality problems and thus help in planning for the launching of planning improvement models. They observe that there are several models, such as Mattson (1992), Sweeney et al. (1997), Oh (1999), all cited in Seth et al. (2005), that incorporate the value construct into quality, as customers do not always buy best quality service but on their assessment of value of service. Most significantly for our purposes they examine all 19 models against the criteria of flexibility to account for the changing nature of customer perceptions and find only 6 models which accommodate that construct.

Walker et al. (2008) argue that the widely quoted Service Profit Chain framework (Hesketh et al. 1970), whilst demonstrating the importance of extrinsic service quality (that perceived by the customer) gives insufficient attention to intrinsic quality (internal quality standards independent of what the customer cannot see). They suggest that the model should be modified to include intrinsic quality and may be particularly important in services high in credence qualities (for example banking). This chapter may show a move in current thinking to align with the perception of quality in the quality literature, suggesting real quality is an alignment of both intrinsic and extrinsic quality.

Svensson (2006) supports the view that there should be a multidisciplinary approach to understanding service encounters and their role in service quality. He argues that the service literature has focused too strongly on the customer perspective and that the construct of service quality is multi-dimensional and current methodologies do not address the inherent complexities and dynamics of service encounters.

Summary and literature conclusions

In summary we conclude that the weight of literature suggests that for real service quality and improvement to be achieved there is a need for integration and interdependence when applying the disciplines and engaging the actors. In practice however there is little evidence to support that this is happening and we find evidence of disparity within the following areas:

- Marketing, quality, and operations
- Intrinsic and extrinsic service development
- Functional, technical, and emotional aspects of services
- Value definition, value development, and value delivery, from the perspective of both external and internal customers.

Further we conclude that popular models used in measuring service quality may not well address the important issues of variety, variability, and flexibility that arise at the moment of the service encounter. The evidence is strong that models such as EBEM, BS, and SERVQUAL are based soundly on empirical studies of critical factors with common ground across major studies. However, there is much less evidence to support the implementation methodologies suggested by these, presenting a need for further research. This presents the real danger that practitioners adopting such models will find that whilst they may be helpful in the diagnosis of improvement areas they are less helpful in developing solutions and in guiding implementation.

Our primary research aims to investigate these issues further and identify what may be critical factors in success and failure for implementing service improvement programmes.

Methodology

The primary research for this chapter involved a survey of bank employees followed by in-depth interviews. The survey took place between March and May 2008 and involved a sample of 224 (from 565 questionnaires issued) bank employees drawn from an internal data base of staff involved in 64 service improvement projects implemented between January 2006 and March 2008.

The survey was organised into the following areas:

- The nature, title, and originating source for the project and the criteria for project selection
- Key performance objectives and performance indicators for each project
- Outcomes
- Other issues of project duration, resources, costs, and general respondent comments and feedback

Following the survey 24 in-depth interviews were conducted from a selection of the completed survey responses. This involved site visits and first-hand observations of improvement projects to gain greater insights.

Tape recordings (audio and video) and transcripts formed the main data for analysis complemented with secondary sources of evidence where available (for example, reports, minutes from meetings, performance statistics, etc.).The data analysis and reduction processes involved the research team and some project participants inductively identifying key themes, data patterns, and associated emotions and behaviours. Respondent checks resulted in follow-up interviews, for clarification of meaning and context. Summary findings were presented back to project groups for comment and to add rigour to the process.

Findings and analysis part one: survey data

This first part presents in summary some of the key findings resulting from the survey.

The nature of projects and originating source

A total of 64 service improvement projects were identified in the period from January 2006 to March 2008. Of these 55 were categorised as Quality Solutions (QS), and originated from the operational areas of quality and/or finance, and involved the use of the EBEM and/or BS. The remaining nine projects were categorised as Marketing Solutions (MS) and originated from marketing and/or sales and involved the use of SERVQUAL or similar customer satisfaction based survey data.

Projects were assessed according to their characteristics and four emerging groups were identified:

- Maintenance – predominantly concerned with regulatory, compliance, or internal administration

- Waste – predominantly concerned with customer complaints, re-work, or inter departmental procedures
- Innovation – predominantly concerned with new product and service development
- Value adding – predominantly concerned with improving existing customer services

The number of projects falling within each category is shown in Table 9.1.

The results show a high degree of bias towards maintenance and waste based projects, with relatively low engagements with customer value adding processes. The results were received with some surprise amongst senior managers, reflecting a dominance of focus not well aligned to the banks central strategic objective to 'build longer term profitable customer relationships through service differentiation'. A possible explanation for this may be that the majority of projects were being driven from specialist departments outside of the main operational areas with some line managers feeling that they had lost control of choice, direction, and control of projects.

Key performance objectives

The key performance objectives indicated by respondents are shown in Table 9.2.

The results show a heavy bias towards organisational cost and organisational control. There is relatively little emphasis on customer satisfaction ratings and very few identified with employee ratings. Many respondents expressed the view that the majority of projects originated from the financial and budgeting control areas, or from quality control. The perception was that marketing and sales were fringe players, and had only a few projects which were largely confined to their own

Table 9.1 Characteristics of service improvement projects

Characteristics of projects	Percent
Maintenance	53
Waste	31
Innovation	4
Value adding	12

Table 9.2 The key performance objectives

Key performance objectives for the project	Percent
Improve customer satisfaction ratings	14
Improve employee satisfaction ratings	4
Reduce costs and improve profit contribution	68
Reduce resources	43
Improve efficiency	46
Improve reporting and internal control	47
Others	12

specialist departments. There appears to be little evidence of integration or interdependence with most projects operating on a departmental basis. Staff confirmed the lack of attention to issues such as job design, roles, skills, and teams, which were largely seen as low priority and a consequence rather than a part of change.

Outcomes

Respondents were asked to rate their overall perceptions of how successful the project had been using a five point scale. The results are displayed in the Table 9.3.

- Mean score = 2.29
- Modal value = 2 (marginally worse than expected)
- Median value = 2 (marginally worse than expected)

The results show disappointing perceptions with only 9 percent achieving a better than expected result and 66 percent achieving below expectations. Some respondents had taken the opportunity to provide additional comments which gave some preliminary indicators of underlying issues, for example, a lack of guidance and understanding, insufficient time and resources, lack of training, and poor project management.

Overall the results were disappointing and warranted further investigation to gain a better understanding of the underlying issues. From the survey results there was evidence of strategic drift in the selection and management of projects, poor understanding and application of the service improvement models used, and low morale amongst staff engaged within the projects.

Table 9.3 Outcomes

Respondents rating based on 5-point Likert scale	No. of responses (n = 224)	Percent
Significantly better than expected (5)	7	3
Marginally better than expected (4)	14	6
As expected (3)	55	25
Marginally worse than expected (2)	110	49
Significantly worse than expected (1)	38	17

Findings and analysis part two: interviews and case study data

In this section we present the results from our interviews, and site visits, using three case study examples. The cases are a Home Loans Centre (HLC) which acts as the main central administrative hub for lending activities, Share Dealing Services (SDS) which provides a specialist investment service for premium customers, and a more general Retail Branch Network (RBN) offering a full range of High Street banking services. The names of the business units are descriptive titles only for the purposes of this chapter and for reasons of confidentiality.

Case 1: Home Loans Centre (HLC)

A key component of the bank's strategy was to organically increase market share in domestic home loans. Aligned with this was the need to grow processing capacity. HLC employs 1800 staff, and its main operation is to take in loan applications and process them through to completion. Loan applications originate directly from customers (personal callers at the office counter, or via the internet), or from originating High Street branch offices located nationally, and a small number from introducing agents (typically small estate agencies). The HLC had been the focus of much consumer complaint over recent years. Under the EBEM process review managers had largely focused their efforts on processing speed; the total elapsed time from the receipt of an enquiry into the issue of a formal offer to lend. Targets and measures concerned time in days from receipt to offer, and increase in overall customer satisfaction ratings. The HLC had successfully over a number of years reduced the average processing time down from 21 days to 95 percent being processed within 3 working days. This was widely benchmarked within the industry as competitive. Yet it was less successful with customer satisfaction ratings which continued to show disappointing downward trends.

> We just couldn't understand it. What's going on? We knew we were better, stats proved it. But then we'd get the BBS [a reference to the Balanced Scorecard performance measure for customer satisfaction] down from HQ and we were down again. (Operations manager)

> Customers never complained direct, not to us, not really. That's why we couldn't understand what HQ was going on about. I suppose we didn't really trust the figures. Some research company I think. Maybe just after more business. (Process worker)

Managers conceded at the outset that they had no real evidence of customer requirements and were unable to understand the declining ratings despite the apparent significant gains in processing times. To better understand this a series of actions were undertaken including an initiative involving managers going back to the floor to gain a thorough understanding of the work from end to end, and front line workers going floor to boardroom to present their views to senior management. The bank also commissioned two major customer research projects, an ethnographic study of families going through the home loan process, and a series of focus groups covering a range of customer profiles.

> It was all a bit scary really. Didn't know what to expect. Thought it was a bit of a gimmick... back to the floor and all that, but it did open my eyes. I could see it wasn't coming together. Not at all. It looked ok on paper but it just wasn't right. Some of the stuff was just plain daft. We'd insist that customers signed a loan acceptance notice... they'd forget and we wouldn't chase them. Except then when they had the removal van outside. Couldn't complete. Lawyers won't allow it. We'd have a panic on. How daft was that. (Loan underwriter)

> It was 3 weeks before they told us. Buyer had pulled out. Found another house. Agent was on holiday. You'd have thought the bank would have told us. (Customer)

A number of significant findings became apparent. The process was built around speed targets which often worked against the overall quality of the service and the total time that the customer had to engage with the organisation. Examples of this occurred at the front line; making appointments with the customer with inappropriate lending officers; the customer not being well briefed on what essential documents might be necessary (causing a delay and second appointment). Loans initiated by branches were a particular problem; branch managers being largely targeted to achieve volume of business; HLC underwriters being targeted on building a low risk lending portfolio.

> This was a constant problem. A battle field. Them and us. We knew they had targets but some of the **** they tried to get through. Just had to send it back. Say sorry. But not having my name on this. (Loan underwriter)

> First we're told we could have it then they said no. Rang the number on the letter and all's they could say was that we'd have to go back to the branch to sort it out. (Customer)

However, the main issue to emerge was that the central performance target i.e. speed of processing was flawed. Customers were not overly concerned with this. Their main priorities (and these were well known to front line workers) were to be 'well managed and looked after' through the process. Indeed timeframes for buying a new house, and completing a house move were accepted as being weeks (possibly months) and not days. Speed of paper work was not an issue. The real problem was knowing what was going on, what happened when and where, and how life arrangements and home moving could be brought together in a sensible plan. The system involved not just the bank and the customer but also, estate agents, solicitors, valuers, insurers, removers, not to mention other families in the inevitable house buying chain. Another key finding was that customers had high expectations that the bank was the key player in managing this process and would consequently tend to apportion most blame to the bank for any failings to warn of a potential problem, even if the problem was largely outside of their control.

Following this review the bank is now looking at addressing the real issues for the customer, has introduced home arranger roles to guide the customer through the process, and changed the organisation structure to localised teams. They believe that this will add value for customers and workers, and help build better managed teams with a closer feel and knowledge of their areas and markets. The early results from this pilot are encouraging and show a decline in service complaints and an increase in staff morale. Further work needs to be done to track the results over a longer period.

Case 2: Share Dealing Services (SDS)

SDS has a relatively small portfolio of clients compared to its main competitors but the business is seen as strategically important to the Group and manages high value customers. Based close to the centre of financial activity it employs over 500 staff, of which it has 130 share dealers. Other staff work in support functions, for example financial experts, economists, legal advisors, and other administrative workers. Over the past three years the nature of the dealers role has changed significantly, partly driven by advanced communications technology, but also a desire to improve the productivity of high cost operational workers. The business had been slowly dismantling dealer teams, in favour of individual dealers with back up from specialised functional support activities. The result was a functional looking organisation structure with separate reporting responsibilities and with support staff housed in separate locations. The process redesigners claimed more efficient use of high value staff, greater flexibility and utilisation of front line dealers, and clearer lines of control and accountability. In practice the dealers were unhappy and complained of isolation, losing client knowledge, over burdened with reporting and controls, and being increasingly pressured to meet unrealistic targets.

> In theory it sounds fine blar de blar, but what these guys don't realise is that you need back up right now. What am I supposed to do when I'm offered a deal and my back up is not picking up his phone, gone off for a fag break, I don't know? (Dealer)

Others complained that the informal discussions within teams were often undervalued, and in fact were a significant part of preparing for dealing.

> I miss the conversation, the banter. Oh it doesn't sound much, but it's surprising what you would pick up. These jobs are all about having your ear to the ground, bits of this and that. You make connections you see. Part of a picture. (Dealer)

> Not managing a portfolio anymore I don't engage with customers, don't know them personal like or their business. The social sides gone with it. Don't know if they're high or dry. (Dealer)

> Difficult to sense what's really going on sometimes. They won't always say, and if you can't see em, you don't know. You don't really know. You don't know if their solid [a reference to being in control of their situation and generally consolidating] or on heat [a

reference to over ambitious trading or risk taking]. (Financial support worker)

An incident well known internally was accounted to us, where a particular dealer had 'lost it and gone on a run' [a reference to serious over-trading and risk taking] which had resulted in a substantial loss and some attention from the regulator.

It was like being down the lane on a sat' day afternoon [a reference to White Hart Lane home ground of Premiership football team Tottenham Hotspur], you've just gone 2 down and your chasing [football speak meaning all your players are attacking]. Then you lose it [the ball], it breaks down [you have no defenders]. As AH would say [a reference to BBC football pundit Alan Hansen] your ****** [isolated at the back]. (Share dealer)

Over the three-year period studied SDS has failed to grow its market share and customer portfolios have generally not performed as well as some of its main competitors. Ironically the principle objective of gaining tighter control of dealers has failed and internally it is accepted that some dealers have been exposed to unacceptable levels of risk in pursuit of individual performance targets. This has forced a re-think and some return to portfolios and team structures is likely.

Case 3: The Retail Branch Network (RBN)

The bank has an extensive branch network spanning the UK with over 400 full branch offices and other agency and sub-branch outlets. Within the industry the trend over the past few years has been customer migration away from branches to self service and internet provision. There have been considerable job losses and branch closures, with the trend likely to continue for some while yet.

For the last three years RBN has been the subject of research to investigate customer satisfaction levels, with an annual survey based around the use of SERVQUAL assessment criteria. A key concern for customers arises within the 'empathy' category where the RBN consistently shows poor results. Typical issues revolve around trust (do you trust your bank to always act in your best interest?), fair play (do you believe your bank has a fair policy on fees and charges?), and relationship building (do you believe your bank rewards you for your loyalty?).

Basically our response has been to tell the staff to be more sympathetic and engage with the customers. At the same time they must meet scorecard targets [reference to Balanced Scorecard performance measurement system] sell more stuff, and get the customer to use the technology more. At the same time you are wondering if you will still have a job tomorrow. (Branch manager)

The RBN response has been investment in staff retraining and development days. The results however have not improved and morale amongst front line workers was described by several managers as very low. Scrutiny of the qualitative data suggests that whilst SERVQUAL has been a useful exploratory method highlighting particular issues, RBN have not developed a depth of understanding (why do customers feel this way?), and consequently have developed only shallow non value adding solutions (what do we need to do?), or developed the ability to handle variety in demand (how do we respond to individuals, in different circumstances, at the point of delivery?). Examples are staff being given unpopular scripted procedures to follow in a standardised and routine way. Along with other similar institutions RBN has suffered criticisms for over selling particular products not well aligned to customer needs. Internally staff feel pressured to hit performance targets and this has resulted in some undesirable work practices.

Discussion of emerging themes

The case studies have helped to identify and confirm some common themes. For example there is confirmation of poor links to strategy and strategic drift, with projects miss-directing resources away from key areas. There is confirmation of a lack of integration of effort between quality, finance, and marketing solutions. Perhaps most significantly in all three cases emerges a failure to understand value from the perspective of the external customer and internal customer, resulting in a focus on miss-aligned development and delivery activities which have led to further errors, complaints and waste management. There is also some emerging evidence from the HLC case in particular, that addressing these issues from the resulting pilot study has led to a marked improvement in performance and has lifted staff morale. Further work is necessary to track results over a longer period, but the evidence so far is encouraging and senior managers are already considering further pilot launches in other business areas.

Table 9.4   Common themes emerging from cases

Strategic issues:	HLC	SDS	RBN
Selection	Poor	Poor	Poor
Alignment	Poor	Poor	Poor
Drift	Weak, drift to emphasis on cost reduction and speed of processing	Weak, drift to short term dividends at expense of long term growth	Weak, drift towards simplistic solution to complex problem
Originating Category	Waste: management of customer complaints	Maintenance: Increase control by centralising experts	Waste: management of customer complaints
Integration of QS / MS	Weak: main focus SERVQUAL	Weak: main focus BS	Weak: main focus SERVQUAL
Value definition	Poor understanding of external customer requirements. No provision for internal customer requirements	Poor understanding of external customer requirements. No provision for internal customer requirements	Poor understanding of external customer requirements. No provision for internal customer requirements
Value development	Poor, focus on wrong things	Poor, focus on wrong things	Poor, focus on wrong things
Value delivery	Poor	Poor	Poor

Table 9.4 illustrates our summary analysis of emerging themes from the three cases presented.

Table 9.5 presents our summary analysis and development work on the principle characteristics evident in quality driven solutions, marketing solutions, and value based solutions. Quality solutions have tended to be chosen by individual departments and have relatively weak links to strategic priorities or customer requirements. The focus of projects has largely been to reduce operating costs. Staff members express concerns that they are burdened with paperwork and reporting as a consequence of monitoring performance measures and formal procedures. Marketing solutions have generally developed from customer satisfaction surveys. However, in-depth knowledge of customer requirements is weak. For example, we found evidence that declining trends were not well followed through to discovery of underlying reasons. Staff express concerns that real issues are not being addressed and that they feel continually pressured to meet sales targets.

Table 9.5 Comparison of critical factors in value, quality, and marketing, and based solutions

	Value solutions	Quality solutions	Marketing solutions
Strategic focus	Strong	Weak	Moderate to strong
Value definition	Strong	Weak	Strong / exploratory
Value development	Strong	Moderate to strong	Weak
Value delivery	Strong	Weak	Weak
Focus / methods	Focus on resonating value for external and internal customers	Focus on: Excellence models, Balanced Scorecard, IS09000, Lean	Focus on: SERVQUAL, Customer Satisfaction Surveys
Integration	Strong through role and organisational adjustment	Weak, tendency to maintain an reinforce status quo	Weak, tendency to maintain an reinforce status quo
Mindset	Service improvement and differentiation through value adding	Builds counter intuitive mindsets	Builds counter intuitive mindsets
Systems thinking	Ability to handle Variety in demand	Tends towards narrow and specialised work silos within command and control structures	Tends towards sales and performance targets within command and control structures

Finally in Table 9.6 we present our summary analysis and development work to suggest a new strategic approach to project selection and management. The approach draws on the empirical data from the survey and interviews. We have found for example, that the majority of initiatives originated from the internal Quality department and had a maintenance focus (53 percent). Whilst internal standards and compliance are clearly important issues, many staff feel that paperwork and monitoring performance has become excessive and a burden. In the waste categories improvement initiatives have largely concentrated on improving the re-work or complaint procedures, rather than tracing back the causes and focusing effort here.

Table 9.6   A proposed strategic approach to project selection and management

Strategic category	Focuses on	Strategy
Maintenance	Security, standards, procedures, control, regulation, administration and record keeping	Be selective, control, focus on compliance not over compliance, employ risk assessment, avoid allowing to dominate resources
Waste	Correcting errors and re-work, complaints handling, just in case records	Shift resources away by eliminating root causes
Value	Define value from both external and internal customer perspectives and develop negotiated flexible solutions	Invest and build. Channel resources. Move skills closer to the market.
Innovation	New product and service development, networks and alliances.	Invest and build. Promote alliances and tap into networks and goodwill.

In contrast value based projects and innovation in products and services has had relatively much less attention.

We conclude that the focus and mix of projects needs to be more carefully controlled to ensure that resources are better directed towards strategic objectives. Table 9.6 below summarises the main categories and suggests a strategic approach.

Summary of conclusions and recommendations

The results are based on a survey and qualitative data from a study of seven independent business units operating within a major bank. Whilst the study has spanned a three-year period it should be noted that particularly harsh environmental conditions are evident within the financial sector at the present time. The results may therefore not be generalised but rather are case specific, presenting some rich empirical evidence to inform academics and practitioners within the context of the UK finance sector, and with possible wider implications for other service sectors.

Our study finds that service improvement initiatives have been dominated in recent years by the use of generic models such as EBEM, BS, and SERVQUAL. Whilst we find strong evidence that the factors within the models are grounded in literature, we conclude that these are largely diagnostic in nature and that the models are weak and not well grounded in areas of methodology and implementation. For example, many initiatives we examined used very simple rating scale measures, and were not well received amongst staff. This led to problems particularly in providing sufficient guidance for practitioners with a resulting tendency for simplistic assessment of problems and weak development of solutions. We found particularly strong evidence of project teams selecting activities for quick gains rather than addressing more complex issues, resulting in a lack of longer term sustainability.

The EBEM and BS are particularly demanding on resources and we find that this over time drains focus and effort away from the main work. Both models have a heavy emphasis on points scoring, target setting, and associated paperwork. We find that this tended to promote short term thinking, competitive behaviour, and in extreme cases exposed individuals to unacceptable levels of risk taking.

All of the models we found tended to promote a silo mentality, with a focus on working within departments and functions. For example, project teams tended to be selected on the basis of department with little cross over. The home loans case was a particular example of this where workers within the main operating department were tasked to improve speed and quality with little consultation with either front line workers or customers. There was little evidence of interdependence and integration of marketing and quality based solutions. Based on these findings we make the following recommendations for further work.

Strategic alignment on customer requirements

Project selection and resource allocation needs to shift away from maintenance and waste categories with a greater priority given to those areas which can add real value to customers. In Table 9.6 we suggest an alternative strategic approach. This may be better achieved by having a deeper understanding of customer requirements than was evident in our case studies. Much of the information we found was very shallow, and predominantly based on simplistic customer satisfaction surveys. These provided some evidence of underlying trends but were generally lacking in depth of understanding and consequently led to the development of weak or inappropriate solutions. A more serious approach may be needed with greater emphasis on qualitative methods and the involvement of front line workers.

Integration of effort

There is a need for greater emphasis on end to end process redesign. This may have implications for organisation structures and individual worker roles. The evidence suggests that structures should be better aligned around the process, rather than seek to maintain functional departmental structures, and that individual roles need to have a higher skill and flexibility content. This suggests that some of current trends we found may be misguided, for example separate front and back office operations, separate enquiry centres, and narrow, repetitive specialist roles. A more serious approach may be needed with a greater emphasis on developing be-spoke solutions.

Understanding value

There is a need to better understand value from the customer's perspective. Methods used were weak and simplistic and a more serious approach is needed. Developing value based solutions is complex and involves a careful alignment of both external and internal staff requirements. Our research suggests that understanding resonating value at the point of delivery is vital, and that this often requires roles that are flexible and skills based.

References

Ackoff, R. (1999) 'A Lifetime of Systems Thinking'. *The Systems Thinker,* June, 1–4.
Ahmed, P. and Rafiq, M. (2003) 'Internal Marketing Issues and Challenges'. *European Journal of Marketing,* 37(9), 1177–1186.
Anderson, J. C., Narus, J. A. and van Rossum, W. (2007) 'Customer Value Propositions in Business Markets'. *Harvard Business Review,* 84(3), 90–99.
Ballantyne, D. (2003) 'A Relationship-mediated Theory of Internal Marketing'. *European Journal of Marketing,* 37(9), 1242–1260.
Berry, L. L. (1981) 'The Employee as Customer'. *Journal of Retail Banking,* 3, March, 25–37.
Berry, L. L. and Parasuraman, A. (1991) *Marketing Services: Competing through Quality.* The Free Press: New York.
Black, S. A. and Porter, L. J. (1996) 'Identification of the Critical Factors of TQM'. *Decision Sciences,* 27, 1–21.
Bossink, B. A. G., Gieskes, J. F. and Pas, T. N. F. (1992) 'Diagnosing Total Quality Management'. *Total Quality Management,* 3(3), 4(1), 223–231 and 5–12.
Bryman, A. and Bell, E. (2006) *Business Research Methods.* Oxford University Press.
Carrillat, F. A., Jarimillo, F. and Mulki, J. P. (2007) 'The Validity of the Servqual and Servperf Scales'. *International Journal of Service Industry Management,* 18(5), 472–490.

Checkland, P. (1997) *Systems Thinking, Systems Practice:A 30 year Retrospective.* Wiley: Oxford.

Davis, T. R. (2001) 'Integrating Internal Marketing with Participative Management'. *Management Decision,* 39(2), 121–132.

de Chernatony, L., Harris, F. and Riley, F. D. (2000) 'Added Value: Its Nature, Roles and Sustainability'. *European Journal of Marketing,* 34(½), 39–56.

Doyle, P. (2001) *Value Based Marketing: Marketing Strategies for Corporate Growth and Shareholder Value.* John Wiley: Chichester.

European Foundation for Quality Management (1990). www.efqm/european-businessexcellencemodel last examined November 2009.

Gronroos, C. (1981) 'Internal Marketing – An Integral Part of Marketing Theory', in Donnelly, J. H. and George W. E. (Eds.) *Marketing of Services* (American Marketing Association Proceedings Series), 236–246.

Gronroos, C. (2007) *Service Marketing Management.* John Wiley: Oxford.

Heskett, J. L., Sasser, W. E. and Schlesinger, L.A. (1997) *The Service Profit Chain: How Leading Companies Link Profit and Growth to Loyalty, Satisfaction and Value.* The Free Press: New York.

Jackson, M. (2003) *Systems Thinking: Creative Holism for Managers.* John Wiley: Chichester.

Kaplan, R. S. and Norton, D. P. (1996) *The Balanced Scorecard.* Harvard Press: USA.

Kotler, P. (2003) *Marketing Management.* Prentice Hall: New York.

Longbottom, D. (2008) 'Real Quality: A New Strategic Approach for Adding Value and Innovation in Services'. *Proceedings of 13th ICIT Conference,* Malaysia (April).

Longbottom, D., Osseo-Assare., E., Chourides, P. and Murphy, W. (2006) 'Real Quality: does the Future of TQM depend on internal marketing?' *Total Quality Management and Business Excellence Journal,* 17(4), 709–733.

MacDonald, M. (2003) *The New Marketing.* Butterworth Heinemann: Oxford.

Mele, C. (2007) 'The Synergic Relationship between TQM and Marketing in Creating Customer Value'. *Managing Service Quality,* 17(3), 240–258.

Morgan, N. A., and Piercy, N. (1998) 'Interactions between Marketing and Quality at the SBU Level: Influences and Outcomes'. *Journal of the Academy of Marketing Science,* 26(3), 190–208.

Normann, R. (1992) *Service Management,* 2nd edition, John Wiley: Oxford.

Oakland, J. S. (2001) *Total Organizational Excellence.* Butterworth-Heinemann: Oxford.

Ohno, T. (1978) *Toyota Production System: Beyond Large Scale Production.* Productivity Press: New York.

Parasuraman, A., Zeithaml, V. A. and Berry, L. L. (1985) 'A Conceptual Model of Service Quality and its Implications for Future Research.' *Journal of Marketing,* 8(4), 41–50.

Piercy, N. (1995) 'Customer Satisfaction and the Internal Market'. *Journal of Marketing Practice: Applied Marketing Science,* 1(1), 22–44.

Piercy, N. (2008) *Market-Led Strategic Change,* 3rd edition. Butterworth Heinemann: Oxford.

Piercy, N. and Morgan, N. (1991) 'Internal Marketing: The Missing Half of the Marketing Programme'. *Long Range Planning,* 24(2), 82–03.

Porter, L. J. and Parker, A. J. (1993) 'Total Quality Management – the critical success factors'. *Total Quality Management,* 4, 13–22.

Saraph, G. V. P., Benson, G. and Schroeder, R. G. (1989) 'An Instrument for Measuring the Critical Factors of Quality Management'. *Decision Sciences,* 20, 810–829.

Sargent, A., and Saadi, A. (1998) 'The Strategic Application of Internal Marketing – An Investigation of UK Banking'. *The International Journal of Bank Marketing,* 16(2), . 66–79.

Seddon, J. (2008) *Systems Thinking in the Public Sector: The Failure of the Reform Regime and a Manifesto for a Better Way.* Triarchy Press: UK.

Seddon, J. and Caulkin, S. (2007) 'Systems Thinking, Lean Production and Action Learning'. *Action Learning Research and Practice,* 4(1), 9–24.

Seth, N., Deshmuckh, S. G. and Vrat, P. (2005) 'Service Quality Models: A Review'. *International Journal of Quality and Reliability Management,* 22(9), 913–949.

Sila, I. and Ibrahimpour, M. (2003) 'Examination and Comparison of the Critical Factors of Total Quality Management (TQM) across Countries'. *International Journal of Production Research,* 41(2), 235–268.

Svensson, G. (2006) 'New Aspects of Research into Service Encounters and Service Quality'. *International Journal of Service Industry Management,* 17(3),. 245–257.

Varey, R. J. and Lewis, B. R. (1999) 'A Broadened Conception of of Internal Marketing'. *European Journal of Marketing,* 33(9/10), 926–945.

Walker, R. H., Johnson, L. W. and Leonard, S. (2006) 'Re-thinking the Conceptualization of Customer Value and Service Quality within the Service Profit Chain'. *Managing Service Quality ,*16(1), 23–26.

Webster, F. E. (1992) 'The Changing Role of Marketing in the Corporation'. *Journal of Marketing,* 56(4), 1–17.

Womack, J., Roos, D. and Jones, D. (1990) *The Machine that Changed the World.* Rawson Associates: New York.

10

Electrifying Performance: A Study of a Systems Thinking Intervention at a UK Electrical Distribution Network Operator

Mark Hopkinson

UK energy companies often attract criticism for the price of energy and poor levels of customer service. This chapter looks at one particular energy company who is trying to improve in order to save money and to improve customer service and customer value for money. The topic of systems thinking is reviewed and one methodology in particular is chosen to apply to the Distribution Network Operator (DNO) arm of the company.

The systems thinking methodology chosen is that of Seddon's and its application to the DNO returns significant improvements in the capability of the system to meet the purpose in customer's terms. It also delivers significant financial benefits for the energy company and an improvement in the measures defined by the industry regulator OFGEM. Whilst Seddon's 'lean systems' methodology has some apparent limitations when compared to other methodologies, and arguably it was not fully applied in this case, it has shown significant improvements and if applied fully, may have realised even greater performance improvements.

Research method

Primary research was conducted through a mixed methods longitudinal study of an intervention which involved the application of Seddon's lean systems methodology to the Repair and Restoration (R&R) activity in a particular DNO. This means that qualitative data was collected using semi structured and unstructured interviews with those involved in the intervention and the R&R activity. Quantitative data relating to the performance of the R&R activity from various stakeholder viewpoints was collected before, during, and after the intervention and was used to determine its success.

An overview of systems theories

Modern systems methodologies stem from the inadequacy seen in reductionist scientific methods when attempting to understand complex systems, as reported by Bertalanffy (1969, 1972). These methods failed to explain common elements seen in systems such as requisite variety (Ashby, 1964), and emergence (Bertalanffy, 1969).

Systems thinking is concerned with 'holism', the theory that whole entities, as fundamental components of reality, have an existence other than as the mere sum of their parts. The Society for General Systems Research has led to diverse research into systems thinking (Flood, 1999, pp. 31–32). Some of the theories and methodologies developed are summarised below.

Systems methodologies

Checkland acknowledges that hard systems methodologies such as systems engineering and systems analysis have been successful in introducing 'systemic rationality' to decision making, where the problem is to select from a number of alternatives in order to reach some predetermined end. However, he believes there are problems when this 'hard systems thinking' is applied to 'soft' problems (Checkland, 1999, pp. 141–144) which is a view also shared by Jackson (2003).

One limitation, mentioned by Jackson (2003) and described by Checkland and Scholes as, 'the Achilles heel of systems engineering' (1990, p. 17), is the reliance on the need being a given. Therefore hard systems methodology such as systems engineering focus on 'how to do it', when 'what to do' is already defined (Checkland and Scholes, 1990, pp. 17). If there are multiple stakeholders, and the specification is unclear, it can therefore be difficult even to get started in many problem situations (Jackson, 2003, pp. 61).

Jackson makes reference to Hoos' (1972) critique of the use of systems analysis and systems engineering to tackle public policy issues in California, as showing the typical problems that arise for hard systems thinking at higher levels of complexity in Boulding's hierarchy (Boulding, 1956). Checkland describes Hoos as, 'very scathing' (1999, p. 142), and suggests the appraisal by Churchman (1968) is a more useful depiction of the limitations.

Alternatives to hard systems

In a response to the problems associated with applying hard systems to 'soft' problems, a number of alternative methodologies were developed, including Beer's Organisational Cybernetics (1980), Ackoff's Interactive Planning (2006), and Checkland's Soft Systems Methodology (SSM) (1990). Seddon has also developed a methodology which he sometimes refers to as 'Systems Thinking' (Seddon, 2005), in this chapter it will be referred to as 'Lean Systems', the name given to the methodology by Jackson (2005).

Seddon's lean systems methodology

'Check Plan Do'

Seddon's approach finds its roots in Japanese manufacturing, and is based on the thinking behind the Toyota Production System (TPS). Seddon does not use the 'lean tools' which many have taken from TPS and applied to other organisations. He has developed his own methods for applying the thinking behind TPS to a service environment.

The approach is centred on the concept of managing the organisation as a system, where the role of management is to work on the system to 'make the work work'. This is the opposite of traditional hierarchical command and control organisations where the decisions and the work are separated, and managers are managing targets and not the system (Seddon, 2005, pp. xi–xvi).

There are three main elements to Seddon's methodology as illustrated in Figure 10.1.

- 'Check': Understanding the organisation as a system, understanding its purpose, capability, and how it operates.
- 'Plan': To understand the levers for change, to determine how the system can be improved so that it better meets its purpose.
- 'Do': To take action on the system, to trial proposed solutions, to monitor their effect on capability, and to fully implement the effective solutions (Seddon, 2005, pp. 101).

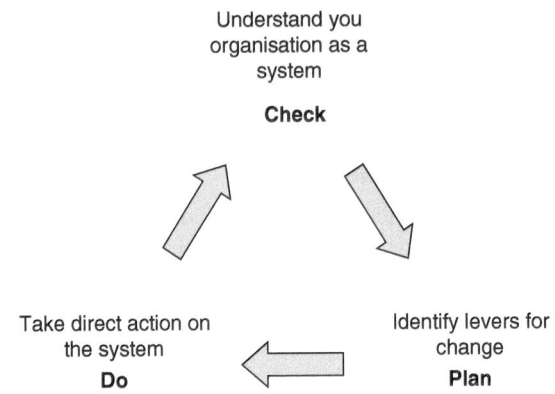

Figure 10.1   Seddon's 'Check Plan Do' methodology
Source: Seddon, 2005, p. 110.

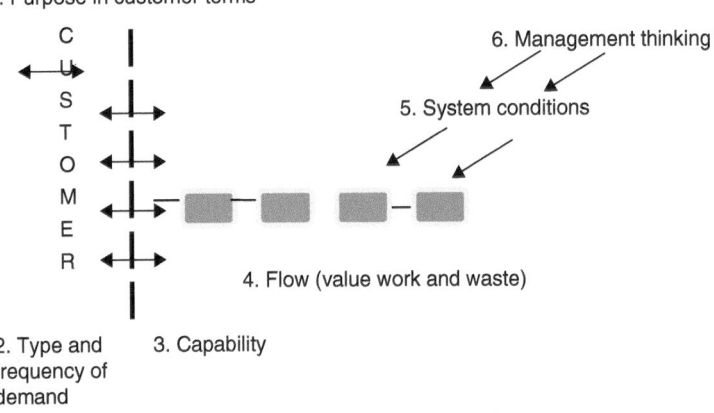

Figure 10.2   The model for 'Check'
Source: Seddon, 2005, pp. 102.

It is important to note that the elements are shown as a continuous cycle, and therefore the methodology is not designed as a 'one off' improvement, more a continuous cycle by which to manage and improve a system. Figure 10.2 shows the model for 'Check'.

The model is sequential; therefore the starting point is to understand the purpose of the system from a customer's point of view. The second step is to understand the type and frequency of demand on the system in order that the demand which is of value from the perspective of the customer, and is high frequency and predictable can be identified. It is then necessary to understand the capability of the system to deal with this demand, and the predictability of the systems' response. Then the flow of work can be studied, and the 'value work' and waste identified. The value work is the work which is necessary to deliver what the customer requires (Seddon, 2005, pp. 108). The final steps are to understand the causes of the waste, in the current system's performance. These are the system conditions, which are often present as a result of management thinking or regulations. Seddon proposes that a systems redesign will create an adaptive organisation capable of meeting changing demands.

Whilst the methodologies of Checkland, Ackoff, and Beer are well documented and reviewed by academics, there is much less published work associated to Seddon's methodology. In fact it is barely mentioned by the literature reviewed so far. It would therefore be prudent to ask, 'is this even a valid systems methodology?'

Jackson (2005), has reviewed Seddon's methodology, and although he lists some limitations such that it concentrates on the purpose from only one perspective, the customer, rather than all stakeholders, he concludes it is a valid systems methodology which has demonstrated positive results where it has been applied (Jackson, 2005, pp. 68–70).

There are similarities between the methodologies of Checkland, Ackoff, Seddon, and Beer. When considering Beer's VSM model, the methodology referred to is that described by Jackson (2003). All are concerned with holism and taking a systemic perspective. All are seeking to understand a problem or 'mess' in a structured way so that it can then be resolved, or 'dissolved'.

Whilst each methodology has been originally defined for their own purpose, the methodologies of Checkland, Ackoff, and Seddon are particularly similar in their undertaking. All follow similar patterns of:

- Understand the system today
- Define some alternatives
- Implement some the changes and experiment

Figure 10.3 is a visual representation of the similarities.

Jackson proposes that systems methodologies have developed from 'hard' systems thinking in two ways. Either they have sought to deal with complexity, or they have sought to deal with pluralism, by which he means a diversity of possible perspectives among those concerned

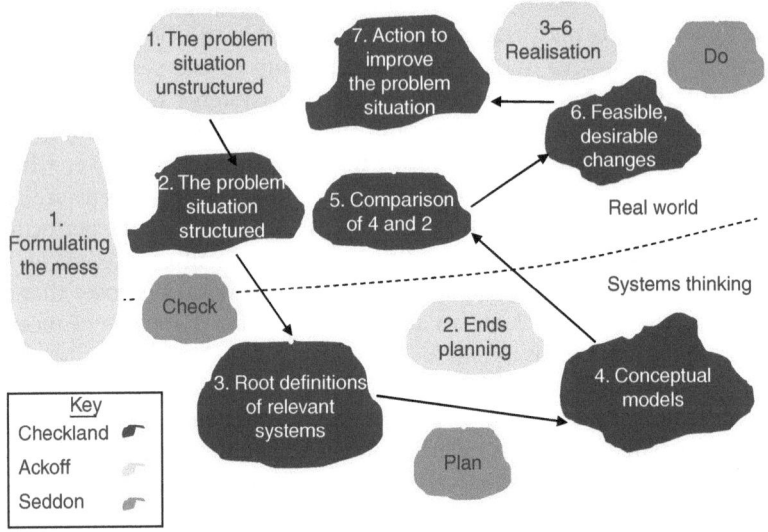

Figure 10.3   Comparison of Checkland, Ackoff, and Seddon

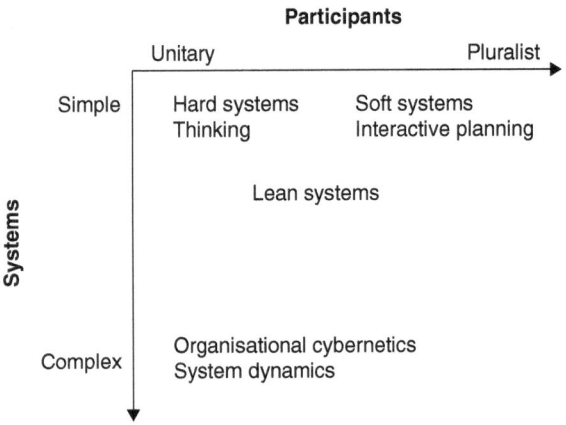

Figure 10.4   The strength of some well-known system approaches
Source: Adapted from Jackson, 2003, p. 24; 2005, p. 71.

(Jackson, 2003, pp. 17–24). Jackson also refers to coercive systems thinking which has not been covered in this chapter. Figure10. 4 illustrates where some of the approaches and methodologies are strongest in terms of their approach to complexity and pluralism.

Figure 10.4 shows that Ackoff's 'interactive planning' is strongest in simple systems, however Ackoff may disagree with this and argue that his methodology is able to deal with more complexity than that of Seddon, due to its ability to encompass all stakeholders. One may argue that this in itself makes a system more complex; however Jackson separates the system complexity from the complexity of stakeholders as shown in Figure 10.4.

Case study background

The UK electricity network consists of over 22,000 km of overhead line, over 1200 km of underground cable, over 1800 major transformers and substations and it supplies power to over 30 million customers (National-Grid, 2008, 2009). The responsibility for operating this network is split between 14 licensed distribution network operators (DNOs). There are also approximately four independent network operators who own and run smaller networks embedded in the DNO networks (OFGEM, 2009).

The Office of the Gas and Electricity Markets (OFGEM) is the organisation responsible for regulating the industry. Customers pay an electricity supplier for the electricity they use, however approximately 20 percent of a typical electricity bill is made up of the charge for transmitting the electricity, which is paid to the DNO by the supplier. OFGEM measures the DNO in a number of ways, including the quality of service, interruptions to supply, and it is also responsible for controlling the price the DNO can charge for transmitting electricity to customers. 'OFGEM, as the industry regulator, administers a price control regime which ensures network operators (DNOs) can, through efficient operation, earn a fair return after capital and operating costs while limiting costs passed on to customers' (OFGEM, 2009).

Repair and restoration

From the money it receives, each DNO is responsible for upgrade, expansion, maintenance, and repair of their part of the UK Electricity Network. One element of this is known as repair and restoration. Repair and restoration is concerned with repairing faults on the electricity network, and restoring electricity where a customer's supply has been interrupted.

This activity involves dealing with calls from customers experiencing a loss of supply, monitoring and controlling the electricity network and where possible providing an alternative supply, providing a temporary supply, and physically repairing the fault. Figure 10.5 gives a simplified

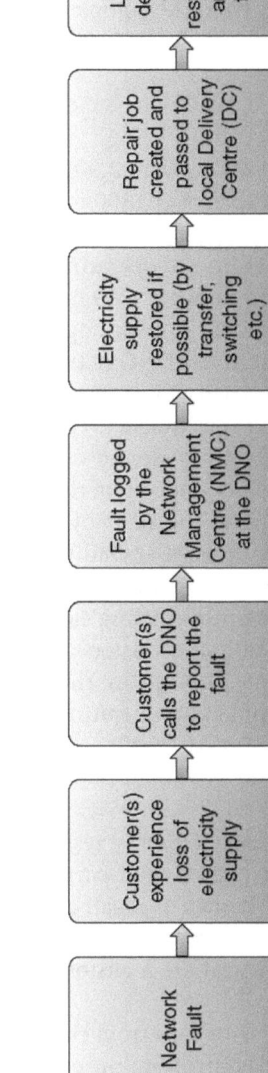

Figure 10.5 Simplified overview of the R&R process

overview of the process. It is the repair and restoration element of the DNO in question which will be the subject of this chapter.

Why change?

The R&R activity is 'reactive', in that it responds to faults on the network; this is in contrast to the planned maintenance and upgrade of the network, which is 'proactive'. If considered from a customer's perspective, ideally there would be no R&R activity, and the network would be 100 percent reliable. If viewed using Womack and Jones's definitions of value and waste (Womack et al. 1990), then a 'purist' may deem the R&R activity to be a waste, in which case the aim would be to eliminate it. In reality, it would be almost impossible and extremely costly to ensure the electricity network is 100 percent reliable, however the DNO in question spent approximately £60 million on R&R activity during 2007, so hoped to reduce this spend, allowing more proactive network maintenance to be completed, thus improving the reliability of the network and the service to the customer.

The performance of the particular DNO's network and its response to faults was resulting in a relatively poor rating from OFGEM, the industry regulator. In addition, when a customer's supply is interrupted, the DNO must pay the customer compensation, depending on the reason for the interruption, and its length. By improving the performance of the R&R activity and the network reliability, the DNO also planned to improve its OFGEM rating and reduce the amount of compensation it had to pay.

System archetypes

Often systems are classified using an archetype, or mixtures of archetypes, such as those listed by Senge (1990). Seddon (2005) also refers to system archetypes, and tailors his approach to suit their differing elements. In his experience of applying Seddon's methodology to many different systems, Davis (2009) defines system archetypes as by what the leverage point is to improve the systems. He defines three archetypes:

- Demand driven—where the main leverage is to pay attention to what comes in to the system and to provide the correct response at first point of contact. In order to predict and improve system performance, understanding type and frequency of demand is necessary.
- Throughput—where what matters is to process the demand through the system as quickly as possible. To improve the system it is necessary

to understand how to process demand right first time, and to study the type and frequency of errors caused by the system.

- Projects—similar to throughput in that the type and frequency of errors are important, but the nature of the project is often each one is new or unique, therefore it is necessary to infer from previous projects. It is often preferable to bring the stakeholders together at the beginning of the project and plan to complete it right first time using the knowledge of previous similar projects.

The R&R activity can be considered as a 'demand driven' system, where the demand coming from the network is breaking down. Seddon (2005) would also refer to it as a 'break-fix' system which can be seen as a subset of 'demand driven'. In terms of Senge's archetypes, it is difficult to speculate how they apply to the R&R activity or the DNO as a whole without understanding the system in more detail. Having identified this system as demand driven, it would suggest that Seddon's lean systems methodology would be well suited to tackle the problems within the R&R system.

Applying 'Check' in the context of R&R service

Having determined the need for change and a proven methodology, small teams were formed from front line operational personnel and management to understand the R&R activity as a system. This was done in two of the seven regional Delivery Centres (DC's). Some of the findings are shown below. It is worth noting that the majority of 'middle management' was sceptical about how the methodology could help, and thought they already understood the R&R system.

Purpose

Calls from customers were studied in the Network Management Centre (NMC) to determine the purpose of the system. This step is critical, as it is the basis for how you assess the current systems capability, and your basis for redesign. It is not unusual, however, to modify the definition of purpose as demand is studied in further detail through Check. In this case one may assume that the purpose is associated with fixing the faults or resolving customer issues quickly. However, the 'Check' team defined the purpose as: 'we maintain a consistent supply of electricity to our customers to keep their lights on'.

So by considering the electricity network as a system, and putting themselves 'in the customers shoes', the 'Check' team determined that the purpose of the R&R activity was to prevent the network from failing in the first place. This made perfect sense; people will have the experience of using electricity daily, taking it for granted that when we switch on a light, it works. We do not want to deal with network faults; we just want electricity available when we need it.

Studying demand

Demand was studied in the NMC, as this is the focal point for all demand into the system. The demand was studied over a three-month period, covering over 12,000 calls. Figure 10.6 shows how the calls were categorised.

They were separated in Value and Failure demand, based on the purpose defined above. Value demand is the demand you want as this is delivering value for the customer, failure demand is the demand due to 'a failure to do something, or to do something right for the customer' (Seddon, 2005, p. 8). Figure 10.7 shows a Pareto of the demand.

Given that the purpose of the system related to keeping customers on supply, any calls caused by the network failing, or a failure to repair a fault 'first time' were categorised as failure demand. It was determined that the only value demand was that caused by someone (other than the DNO) damaging the network and requiring the DNO to repair it, or when a customer wants to cancel some repair activity it may have been resolved another way (resetting a trip switch, etc.). Therefore the team concluded that only 2.35 percent of the demand was value demand, the remaining 97.65 percent was failure demand.

	%			%	
I have no supply	42	Failure	I have an issue with the service received	1.5	Failure
My supply is still off	14	Failure	Can you fix....	1.2	Failure
I have a question for my supplier	13	Failure	There is a flicker on....	1.2	Failure
I have a question for another department	9	Failure	Can you re-install....	0.9	Failure
I have a Mpan (electricity meter reference) enquiry	8	Failure	Can you cancel...	0.4	Value
I am calling from out of the area	5%	Failure	I have had an accident	0.03%	Failure
Can you check...(safety)	3%	Failure	I have damaged your cable	2%	Value

Figure 10.6 Summary of demand in the NMC

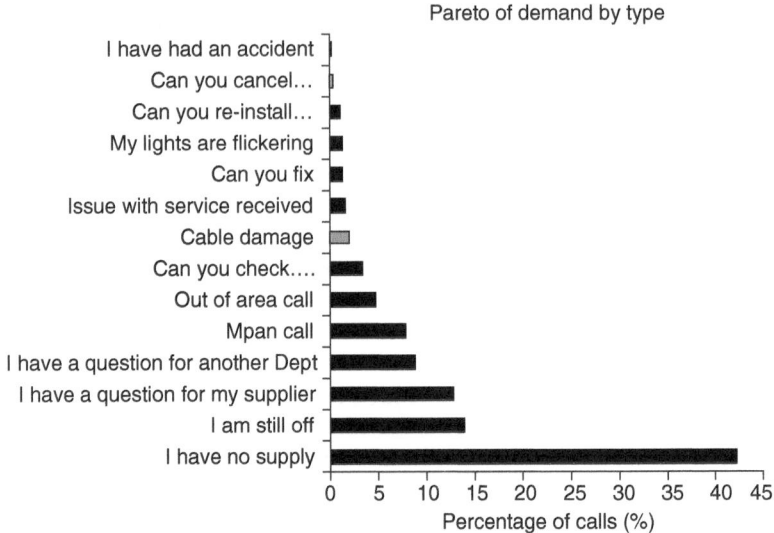

Figure 10.7   Pareto of NMC demand

Capability

Capability measures show how well the system responds to demand, and are derived from purpose. They are leading measures and by understanding and acting to improve them, the system will be improved (Seddon, 2005).

The measures associated with the R&R activity with the DNO were generally speaking 'lagging measures'. They were mainly financial measures based on operating the function within budget, and performance measures based on the regulator's (OFGEM) criteria. Whilst it may be argued that they gave an indication of the performance of the function, this was the performance in the past, and the data was of little use in understanding how to improve in the future. Seddon goes further and warns that acting on lagging indicators causes waste and sub-optimises the system (Seddon, 2005, pp. 49).

The 'Check' team therefore determined four measures based on the purpose, to help understand and improve the system. These were:

1. Vi: Volume In, the volume of faults received by the system
2. Vo: Volume Out, the volume of faults closed or resolved by the system

3. TET: Total Elapsed Time, the time it takes from receipt to closure
4. FTF: First Time Fix, how many of the faults are fixed the first time (and therefore how many do we have to revisit as the initial fix was inadequate)

Understanding the volume of faults in and out of the system enables better capacity planning which will be explained further in redesign; however it also gives an understanding of whether the overall system performance is improving. If we refer back to the purpose defined earlier, it is clear that the customer would prefer the number of faults received by the system to ideally be zero. By understanding this measure over time, it is possible to determine whether actions to change the system have a positive or negative overall effect.

If the electricity does fail, the customer wants the fault fixed quickly and disruption minimised by not having follow-up visits and power interruptions, hence the measures of total elapsed time and first time fix.

Flow

The team then mapped the flow by walking through the process and following what happened to the demand. They understood what happened at each step, and by referring to the purpose were able to understand if the step was value, work, or waste. They also documented any issues which would have a negative effect on the four capability measures described above. In the first instance only two of the regional DC's were visited, but these were chosen in such a way that the majority of issues found would be common across all DC's. Figure 10.8 is the author's interpretation of the flow and the key findings, constructed from the data collected and from talking to those involved in 'Check'.

The process essentially begins with either a customer complaining about a fault, or by the NMC being made aware of a fault through their monitoring of the network. As some of the network can be controlled remotely, an attempt is made to rectify the fault if possible. It was seen that customers were making multiple calls as they received insufficient information about the status of a fault, and also that the details collected by the NMC were incomplete.

The fault then would be converted into a work instruction or job card and a team despatched. During this process, there would be confusion over ownership of the fault between the NMC and the DC based project manager (PM), and there was also excessive paperwork and repetitive checking.

The task of organising and dispatching a team was difficult due to the poor information from the NMC, and also the number of parties involved. This resulted in delays and waiting on site, which ultimately extended the length of the fault and increased overtime.

Due to these delays and missing materials, often a temporary fix is performed so that the customer's power is restored. A repeat visit is then required to fix the fault permanently. Due to the volume of faults and the length of them, there is a backlog of faults requiring a permanent fix. Whilst the team had uncovered a number of issues, it is critical that the root causes of these issues are understood. Seddon refers to these as system conditions.

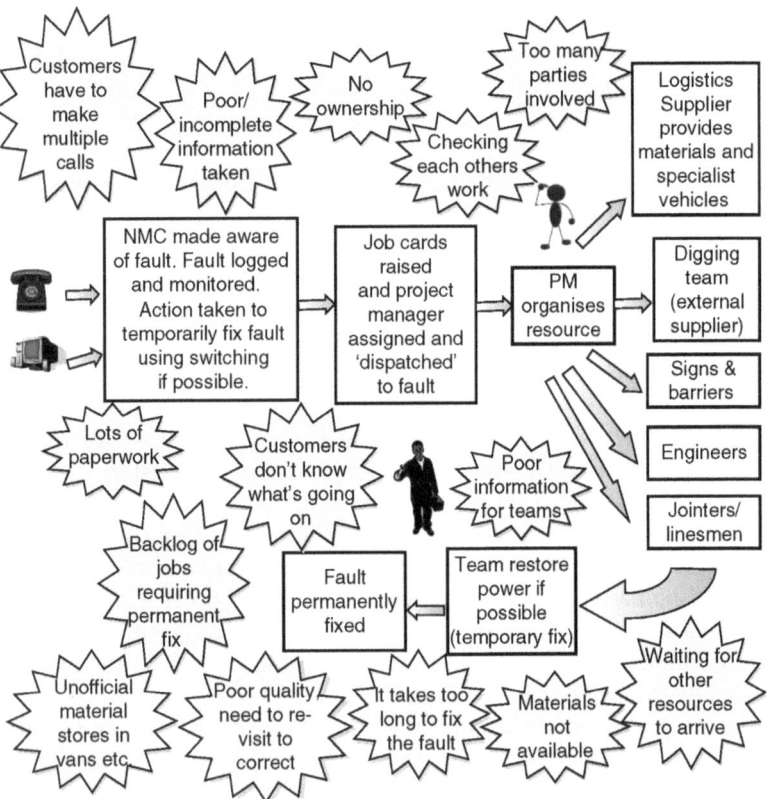

Figure 10.8 Overview of the 'Flow'

System conditions and management thinking

System conditions are the things which explain why the system behaves in a certain way; as such they are also the causes of waste, particularly in a command and control environment. Bicheno lists some typical system conditions as 'structure, processes and measures' (2008, p. 23). Seddon admits that 'the factors influencing behaviour in organisations are many and complex', and include structure, policy, procedures, and measurement (2005, p. 108). Seddon believes that system conditions are created by how managers think about the design and management of work (Seddon, 2005). This is demonstrated in Figure 10.9, showing Seddon's belief that thinking governs performance.

Seddon refers to a Japanese guru, who went to great lengths to make the point to his board that: 'unless you change the way you think, your system will not change and therefore the performance won't change either' (Seddon, 2005 p. xiii). This is a key point, and is critical to Seddon's methodology. Management thinking needs to be changed from 'command and control' to 'systems thinking', if major improvements are to be realised. Below are some of the system conditions highlighted by the team, and an explanation of how they cause some of the issues shown in Figure 10.8.

Measurement

Seddon believes this is the most important of all system conditions (Seddon, 2005 p. 110). Bicheno (2008) agrees and cites Spitzer (2007) as making a strong case that measures are the most important force influencing the culture of an organisation. In relation to targets, Seddon believes they cause problems such as increased variability, waste, demoralisation of the workforce, and ultimately in people finding ways to 'cheat' to meet the target, rather than focusing on the work (Seddon, 2005).

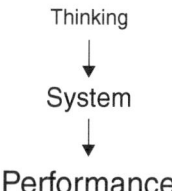

Figure 10.9 Thinking governs performance
Source: Seddon, 2005, p. xiii.

In the DNO in question, measures surrounding the R&R activity were focused in two areas, as described previously: financial and regulatory. The financial measures were basically associated with monitoring the performance of the operation against budget, 'is it costing more to run than we planned, and if so why'. This caused the managers to focus their attention on monitoring financial performance, and not on improving the work. The 'obsession' with performance against budget is flowed down the supply chain, with the suppliers of logistics and sub-contract labour all focused on managing activity and resource to cost, rather than understanding the work and improving the system. It stems from fundamental assumptions behind management thinking that budgets bring control and that targets improve performance.

In the NMC, the focus is the call time; the industry watchdog OFGEM monitors each DNO on its 'quality of telephone response', by surveying customers monthly. Each DNO is scored in the following areas:

- The politeness of the members of staff
- Their willingness to help
- The accuracy of the information given (if information was given)
- The usefulness of the information given (if information was given)
- The speed of telephone response (OFGEM, 2008, p. 15).

Each year the mean scores are calculated and the DNOs are ranked in performance; the best receiving a financial reward, the worst receiving a financial penalty. *It is therefore logical to conclude that the NMC is more concerned with answering the calls quickly and politely than it is with the quality of data it records for use by the team tasked with repairing the fault. This leads to poor or incomplete information about the fault being passed to the team, resulting in waste.* Other measures OFGEM uses to rank DNOs include a CI and CML (OFGEM, 2008, p. 52):

- CI's: The number of customers interrupted per year. The number of customers whose supplies have been interrupted per 100 customers per year over all incidents, where an interruption of supply lasts for three minutes or longer, excluding re-interruptions to the supply of customers.
- CML's: The duration of interruptions to supply per year measured by the average customer minutes lost per customer, per year, where an interruption of supply to customer(s) lasts three minutes or longer.

There is also a compensation scheme for customers who are without power for more than 18 consecutive hours. Therefore there is a focus at an operational level to restore supply as quickly as possible, as this reduces the CML's and the compensation paid for supply being off for over 18 hours. *There is little incentive for the fault to be completely repaired and reinstated to an acceptable standard, hence the backlog of faults requiring a permanent fix, and the repeated visits to site to correct 'quality problems'.*

Overtime

The problems described above, inevitably lead to jobs over running. Whilst this has a negative impact on the customer, it can have a positive impact on the operational personnel assigned to the fault. They are paid overtime at a premium rate, while they wait for materials or barriers etc., so there is little incentive for them to improve the system and reduce this time. Also the system has been operating like this for sufficient time that this overtime has become 'guaranteed' and many of the workforce rely on it. This issue needed to be tackled head on during redesign, if the system was to be improved.

Organisational structure

Generally the R&R organisation is split by skill type, so that all the call handlers are located together in the NMC, which is usually a long way from where the fault occurred. The jointers and linesmen are seen as a key skill and are retained in house by the DNO, logistics, and civil work is outsourced for flexibility and economies of scale. This all leads to many handoffs between all the parties, leading to errors and waste. It also makes it difficult to get accurate information about the status of the fault to the customer. The management thinking here is that splitting by function leads to efficiency.

As a result of management thinking, decision making was made by managers, away from the work, and without meaningful data (capability data relating to purpose for example). This is evident as if managers had made decisions based on a thorough understanding of the work and of the capability measures, the issues highlighted above would not have been present. This led to a divide between management and front line personnel as the sub-optimised system led to frustrations and decisions were made which to those in the work seemed not to make sense. This is demonstrated by the results of a survey conducted as part of the intervention which showed the feelings of management and front line personnel toward each other and toward the company. The survey also showed that all agreed the system needed improving.

Summary of 'Check'

It was clear from 'Check' that the R&R system at the DNO in question was far from optimised, and the combination of system conditions imposed by the DNO management as a result of their 'thinking', and from the regulator were having a negative impact on the performance of the system to meet its purpose as defined above. The result was a 'de facto purpose' which was defined by the 'Check' team as: 'meet regulator targets and make a profit'; with measures against target, against budget, and against competitors in the form of the OFGEM league tables.

What the 'Check' team discovered was that the purpose should be along the lines of: 'We maintain a consistent supply of electricity to our customers to keep their lights on'; with measures of capability defined from the purpose.

Redesign

After presenting the findings from 'Check' to senior management within the DNO, the 'Check' team secured support to redesign the R&R system in all the regional DCs. To this end a 'Lean Academy' was formed consisting of members of the 'Check' team, who were now taken out of the 'day job' and permanently challenged with redesigning the R&R system. One key objective of the Lean Academy was to share their knowledge with people in the work, to enable them to understand R&R as a system and to be able to make improvements to the system themselves, by following Seddon's methodology. As redesign involved the Lean Academy, as full-time team members, and other personnel from R&R and related suppliers and departments, those involved in the redesign will simply be referred to as 'the team'. It is worth noting that following this presentation sceptics amongst management reduced, and some of the initial sceptics who were involved in 'Check' actually became the biggest advocates and were made part of the Lean Academy.

Principles and method

Redesign involves moving from 'Check' to the 'Plan' and 'Do' stages of Seddon's model shown in Figure 10.1. Although the steps are shown as separate, in reality they can overlap, and this is demonstrated in the following below.

Plan

In 'Plan' the team explores potential solutions to eliminate waste and uses the following questions as a framework to determine what the purpose of the system should be and how it can be improved to meet it (Jackson et al. 2008, p. 188):

- What is the purpose of the system from the customer's perspective?
- What needs to change to improve performance against purpose?
- What measures are necessary to gauge improvement?

In Checkland's SSM methodology, this would be termed as making and testing 'conceptual models', where the models are created without real world constraints. Ackoff has a similar stage in his methodology, referred to as 'ends planning', defining an ideal state as if it were possible to start from scratch tomorrow. Whilst Seddon's methodology may not go quite as far, it is encouraged in practice for the team to start with only those steps of the flow deemed to be of value from a customer's perspective and work up a realistic solution from there.

In this particular case the purpose was defined previously as: 'We maintain a consistent supply of electricity to our customers to keep their lights on'. The team added to this by considering that whilst the long term goal should be to have an electricity network which was 100 percent reliable, in reality it was likely to have faults in the short to medium term. They agreed on the following criteria that customers wanted from the system:

- Keep my lights on
- If they go off, get them back on quickly
- Do not switch me off later to complete work
- Keep me informed

In terms of what needs to change, the team identified that the whole R&R system needed to be redesigned, and that this would need to include the suppliers of civil resource, sign and barriers, and logistics and materials.

The measures were the capability measures described above.

Do

In 'Do', the team implement solutions gradually, by experimenting and testing their ideas, and seek to consider further improvements. 'Do' has the following key elements (Jackson et al. 2008, p. 188):

- Develop redesigns with those doing the work
- Experiment gradually
- Continue to review changes
- Work with managers on their changing role

In the 'Do' stage it is important to involve those doing the work. They have valuable knowledge about the system, and their involvement in the redesign helps to break down the barriers in the organisation, which in this case were evident from the staff survey. *Experimenting allows for different ideas and solutions to be tested, and in conjunction with capability measures (related to purpose) the effect of these solutions on the performance of the system can be judged with data, rather than opinion.*

A key element of 'Do' is to work with managers to change their role, moving away from managing resource, to the management of end-to-end flows. From managing people to acting on the system (Seddon, 2005). As discussed before, management thinking also needs to be changed, as it is this thinking which leads to many of the system conditions which sub-optimise the system performance.

This stage is aligned to the real world development and implementation of feasible and desirable changes in SSM, or the realisation stages of Interactive Planning. All three methodologies advocate implementing a process as close to 'ideal' as possible, and recognise that experimentation, or adaptation of the conceptual models may be required. In 'interactive planning' Ackoff focuses on determining the resource required and determining and action plan, whilst this is not mentioned in Seddon's methodology, it is implied and undertaken in practice.

Based on their learning from 'Check' the team defined some operational principles to be used in redesign:

- Purpose must relate to 'what matters' to customers
- Measures must also relate to 'what matters' to customers
- Decision making within the work
- Minimise hand-offs
- Ensure that staff have the skills and knowledge to pass on work clean
- Focus on and strive to only do the 'value work'
- Learn and improve
- Change based on accurate, relevant data not gut feel

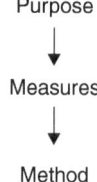

Purpose

Measures

Method

Figure 10.10 Fundamental principles of work design
Source: Seddon, 2005, p. 49.

The team agreed to start with one of the DCs where they originally performed 'Check', and would involve as many people from the R&R team at that DC in redesigning the system. The method of redesign followed Seddon's principles of work design as shown in Figure 10.10.

This means that the purpose should be defined from the customers' point of view. From this purpose capability measures should be defined to determine how effective the system is at meeting the purpose. This leads to a method being defined which best delivers the purpose, as determined by the capability measures.

Redesign in practice

A System for measurement

The team had a defined purpose and a set of capability measures relating to the purpose, however the measures were not in place, and therefore the result of any changes to the system could not be judged. The team had to develop a measuring and reporting system to enable them to understand the capability of the system, and so the Online Measures Application Program (OMAP) was born. It allowed anyone within the R&R organisation, who had access to the company intranet, to interrogate data relating to the volume of faults received and cleared, and the total elapsed times for the faults. At the time of writing this chapter, the first time fix rate still had not been implemented.

The data in OMAP was the basis of the team's decision making in how to redesign the system.

A design to minimise waste

One of the significant areas of waste highlighted in 'Check', which affects the faults total elapsed time, was travelling between faults. Each DC had an area, loosely based on counties within the DNO in question's region. It was not uncommon from resource to have to travel across the county on a daily basis, attending different faults. This led to a concept of geographic 'patches' within each DC with dedicated resource, so that travel time was minimised. The data in the Online Measures Application Program (OMAP) provided a vital insight into how these patches could be designed.

Figure 10.11 and 10.12 show the type of work received per hour. The sample is an average for the month of October 2008. The different types of work are shown by the different coloured blocks. It is clear that the type of work varies between the two areas, this is due to a number of factors, the main one being that area A is a large urban conurbation, and area B is rural. The main differences between the type of work in rural and urban areas are:

- In rural areas the majority of the cable is 'overhead', between poles and pylons. In urban areas it is predominantly underground, buried under pavements and roadways. Engineers and craft teams tend to specialise in either overhead lines or underground cables, therefore a different skill set would be required in each of the two areas.
- Work in rural areas tends to require long distance travel, whereas in urban areas travel to faults is usually shorter, but the time to reach them depends heavily on the traffic at the time of the fault. This is therefore an important factor to consider when planning.

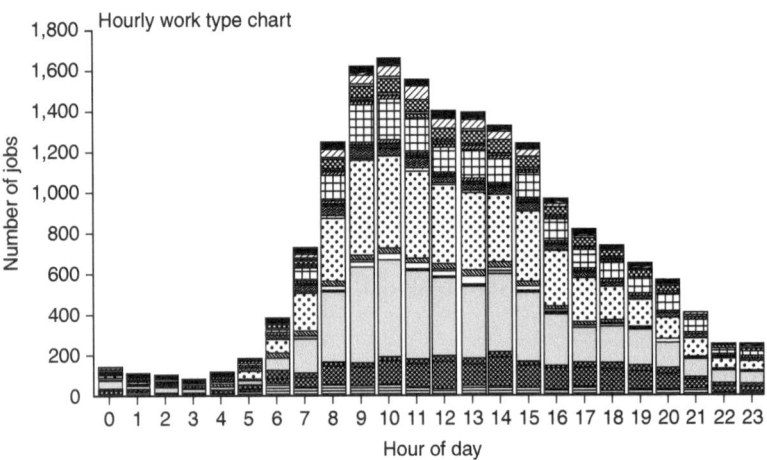

Figure 10.11   Work type in area A

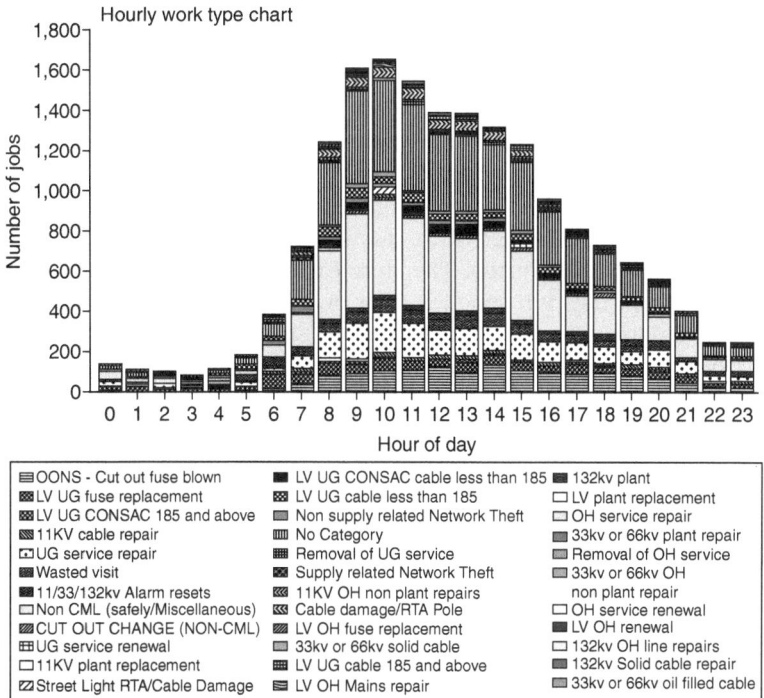

Figure 10.12 Work type in area B

The shape of the graph also provides an insight, as it shows that the vast majority of calls are between the hours of 08:00 and 20:00, which is when most people are awake and active. If a fault occurred in the early hours of the morning, it is likely to go unnoticed as most people would be asleep, and so would not report it until they wake up.

What is important about this data is that it is predictable; that is to say it is predictable that in area A the majority of faults will be with underground cable, and it is predictable that the majority of faults will be reported during the times shown. It is therefore possible to design the patches against this demand.

The team analysed this data for the DC, and split the DC into a number of patches, based on geography, work type and volume of faults. In this case the first DC was split into six patches.

Resource and roles

Each of the six patches had dedicated resource attached to it, determined from the OMAP data. This included underground jointers or overhead linesmen for example. Also the core hours that the majority of resource was available was also dictated by the OMAP data. It was mentioned previously that overtime was an issue. During the redesign, operational personnel were involved, and as a result of them understanding the demand data they agreed with management that overtime should be reduced to benefit the overall system.

Instead of having different organisations interacting with each other, teams were formed with resource from the various organisations assigned to the patch. Therefore each patch got designated civil resource from the civil supplier, their own vehicles, sign and barriers, and materials stores. The materials in the stores were chosen based on the demand from the network in each patch, this reduced inventory by over 20 percent, but ensured the materials which were needed were available. Eliminating unofficial stores was arguably the biggest inventory saving, but this was very hard to quantify due to the nature of unofficial stores.

In addition to the operational resource, each patch was assigned a patch manager. The role of the patch manager was to coordinate resource in the patch using the measures in OMAP. In reality, this role was split between the existing project managers, depending on who is working at any particular time. Area managers were also created, who were responsible for a number of patches; in the first DC there were three area managers, each with two patches. The role of the area manager was to understand what was stopping the work from happening in the patches (the flow) and taking actions to improve it. A role called 'Quality of Service Today' was also created to remove waste in the system, in reality this ended up being combined with the area manager role. The area managers were normally encouraged to highlight and work on their 'top five wastes'. A 'Quality of Service Tomorrow' role was also created, and there was generally one per DC. This role involved looking at longer term improvements; an example would be that the area managers may be reporting that 'joint type X always fails after two years', so the QOS tomorrow person would take this information to the people responsible for designing the network, and the procurement department, and would get the joint changed or improved. This would improve the reliability of the network, and would ultimately reduce the number of faults, thus improving performance to the customer.

The NMC is obviously an important part of the system, so work was done to improve the quality of data passed to the patches. The NMC worked with the patches to determine rules for the different types of faults. For example for a type 1, the NMC always needs to seek advice from the patch manager before passing the work on, whereas for a type 2, the NMC can despatch a team without contacting the patch manager. Also rules were dependant on the time of day, so in the first DC, the NMC was responsible for dispatching the teams to the faults from 19:00 until 07:00 when volumes are low. This reduces the amount of resource required in the DC, or on standby. When the demand is greatest, the DC takes control of dispatching the teams, as they are closer to the work.

The team developed a set of rules for everyone involved in the system to work to, these were:

1. Decide on data (no data, no can-do)
2. Design against demand, always!
3. Do work perfectly first time (fix)
4. Own the process (closed loop)
5. Set measures and use them (be visible)
6. Identify and act on causes of waste
7. Consider the customer when making decisions (act in their favour)
8. Have fun!
9. Always walk your crops everyday

These rules allowed people to challenge behaviour, for example 'why are you sitting behind this desk and not out in the work?', 'where's the data that supports your proposal?' Figure 10.13 is a pictorial representation of the DC design used by the team in the first DC. It shows the six patches, and the various roles and resources in the centre. In the top left hand corner are the rules, in the top right are the rules of engagement for the NMC and the DC. In the bottom left are the measures, and the vision 'moving from fire fighting to preventative maintenance, leading to switching off work'. In essence reducing faults by effectively repairing and improving the network. As described previously, redesign involves experimenting, and there were numerous iterations and improvements before this design was agreed upon; these are not covered in this chapter.

Roll out vs. Roll in

Once the new design was in place at the first DC, it was clear from the OMAP data that the design was having a positive effect on performance of the system. Morale improved as frustrations were removed, and costs fell as waste was taken out of the flow. The next task therefore was to improve the system in the other regional DCs.

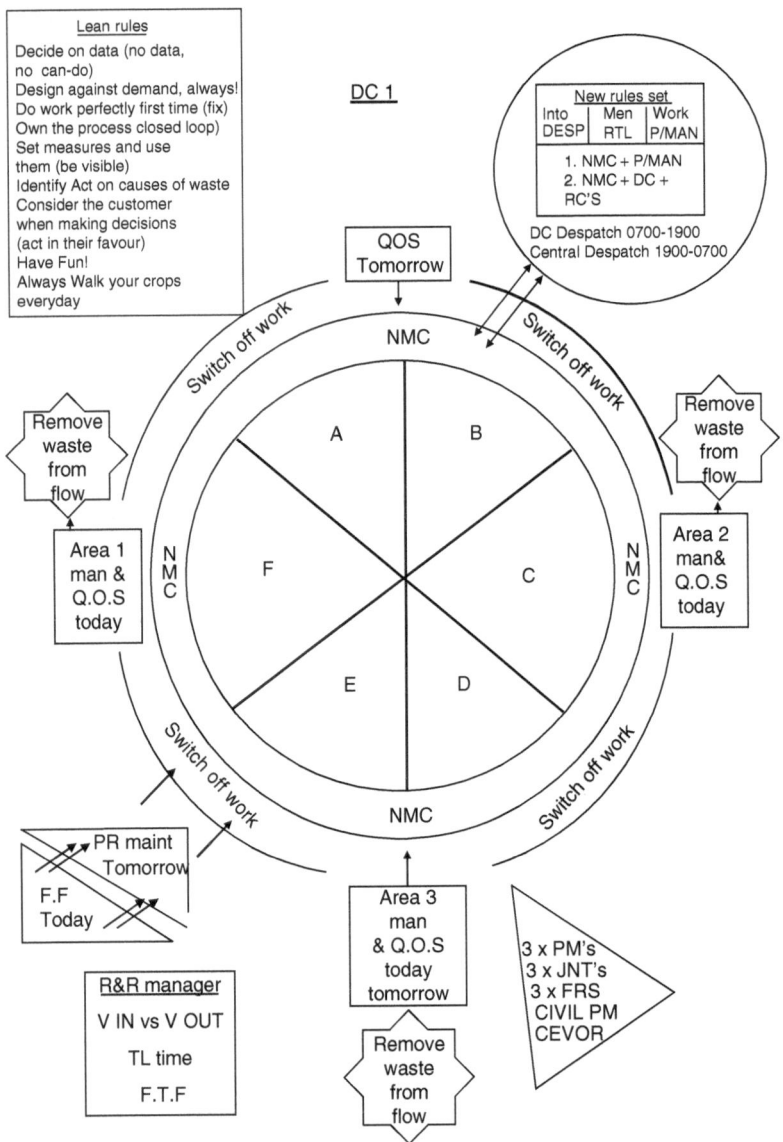

Figure 10.13 DC design example

Traditional command and control thinkers would be tempted to 'roll out' the solution; in the extreme this would mean taking the model from the first DC and imposing it on the other DCs. One major disadvantage of this method is that there would be no learning in the other DCs, and as a result thinking would not change. It has been established earlier that thinking needs to change for the change to be successful. The alternative approach used by the team, was to 'roll in'.

Employees from the various DCs were invited to come to the first DC and learn about how to understand the system using 'Check' and how and why the system was redesigned. Using the principles they would then go and understand the R&R activity as a system within their DC, and use the data to design their own patches. As a result, individual DC's would determine the number of patches, and levels of resource, based on the demand data for their region. This meant that each DC understood their demand, and were fully bought into their design for that region.

Senior management support for the activity also increased as positive results were seen. When visiting the DCs, it was not uncommon for senior managers to ask 'show me your top five wastes', thus showing a level of understanding and support to the staff in the DCs.

Results

There are undoubtedly improvements in morale and the overall working environment, which can be hard to quantify. Ultimately, however, the success of the redesign would be judged on quantifiable results. There are a number of stakeholders involved, and each would judge the success differently. Those closest to the work would use the capability measures defined by the team during 'Check', as these show how the system performs against purpose. Senior managers and company shareholders would be more likely focused on cost: 'how has the redesign improved the bottom line?' OFGEM, the regulator, would be looking at the measures defined in their quality of service report, CI, CML and the quality of telephone response.

Each of these viewpoints will be considered when assessing the quantifiable results of the change.

Capability performance
The capability measures introduced by the team measured the volume of faults into the system, the volume cleared, and the total elapsed time for the faults. It was not possible to reliably measure the first time fix

percentage. All the following control charts are individual charts (X charts or run charts) and were interpreted using guidance from Wheeler, 1993, and Bicheno et al. 2005.

Figure 10.14 is a control chart showing the volume of faults received per day for all DCs. The data is taken from July 2007 to July 2008, and the data has been split at July 2008 to allow a comparison of the performance between 2007 and 2008 and between 2008 and 2009. The data shows that the mean number of faults decreased from 342 to 293, between 2007 and 2008 and between 2008 and 2009, a reduction of 14 percent. It also shows that the control limits have narrowed, indicating the process is more 'in control', and that the number of special causes (those outside the control limits) has reduced from 6 to 3. Figure 10.15, shows the volumes out. This follows a similar pattern with the mean number of faults received reduced by 12 percent from 324 to 282 and the control limits are tightened. This is to be expected as clearly the volume of faults received and the volume cleared are strongly linked.

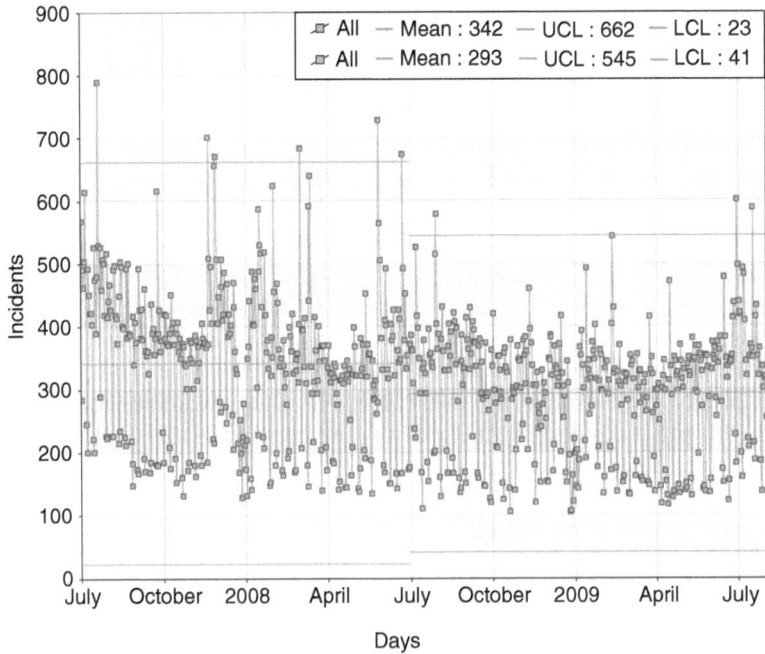

Figure 10.14   Control chart of volumes in

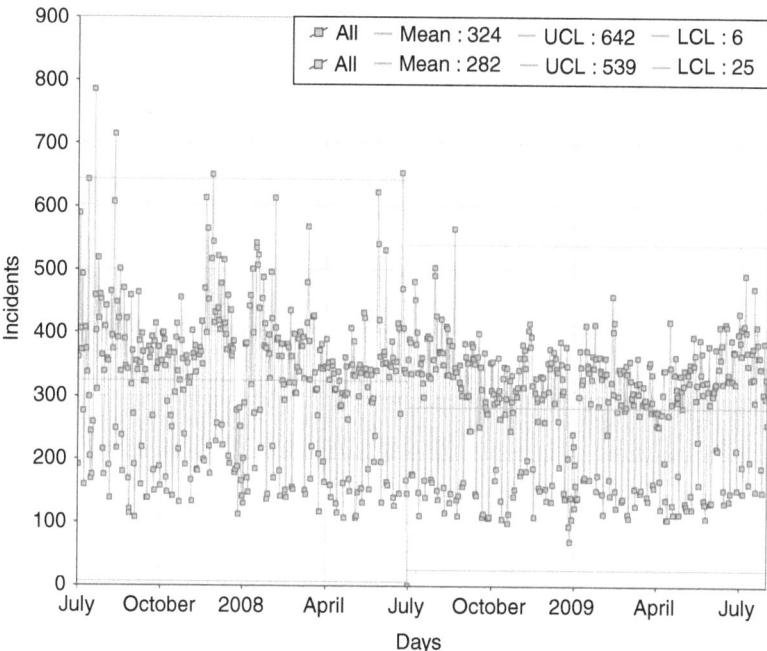

Figure 10.15 Control chart of volumes out

This level of improvement is impressive given that the main factors influencing the volumes of faults are the improvements to the network which can take a long time to implement and impact the figures.

Figure 10.16 is a control chart showing the total elapsed time (TET) for the same period. It shows the mean TET has fallen from 8,135 minutes (5.6 days) to 5,725 minutes (4 days) from 2007–2008 and 2008–2009, a reduction of 30 percent. The process is also more 'in control' as the control limits have significantly tightened. If someone asked for an estimate of the TET for the next fault, the answer based on the data would be the upper control limit, as this statistically is the upper limit of the process. A promise of the mean value would result in a failure of the promise at least 50 percent of the time. Therefore the expected fault duration that the DNO could quote to the customer has fallen from 28,622 minutes (20 days) to 15,970 minutes (11 days), a reduction of 44 percent.

This clearly shows that from 2007–2008 and 2008–2009, the capability of the process in terms of faults received and cleared, and fault TET

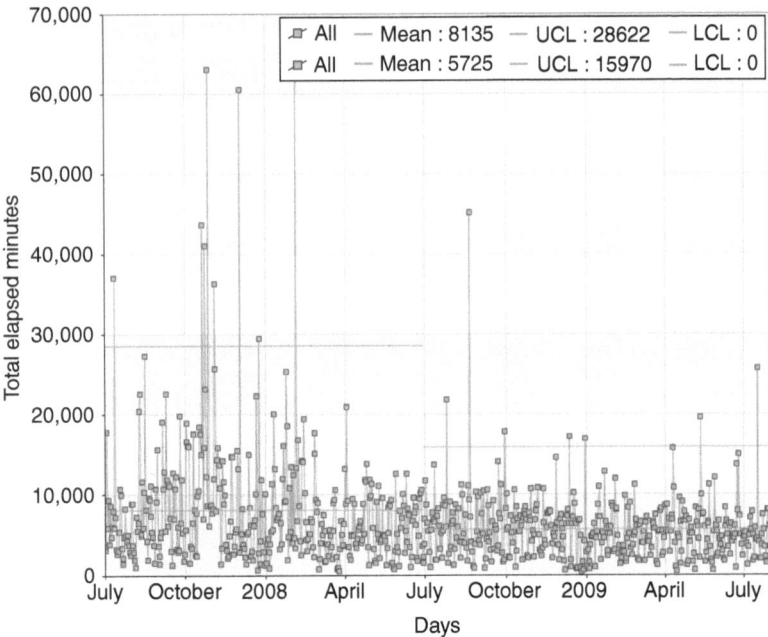

Figure 10.16 Control chart of total elapsed time

has significantly improved. As these measures relate to purpose in the customer's term, it is fair to surmise that the performance from a customer's perspective has improved.

Financial performance

There was a significant cost to developing and implementing the improved model, as summarised below. However, the savings generated return this investment many times over.

The cost to develop and implement the model in all the regional DCs was approximately £2 million. With additional costs of around £600k in the first year of implementation and £900k every year going forward. Therefore, costs in 2008 were approximately £2.6 million, with estimated ongoing costs of £900k per year.

To secure the investment the benefit was predicted to be £6.5 million per year, when compared to performance in 2007. Clearly this would justify the initial project cost, and the additional costs going forward.

The actual savings achieved in 2008 were £14.5 million; these are actual bottom line savings and not projected or estimated figures.

Some significant savings included £6 million due to the reduction in customer minutes lost (CML), £2 million reduction in overtime and £1.5 million in stores and vehicle costs.

Performance against OFGEM measures

As discussed previously, OFGEM, the industry regulator measures the DNOs using a number of metrics. The 'quality of telephone response' measure is subjective and its value is questioned by some at the DNO in question. The more useful OFGEM measures are Customer Incidents (CI) and Customer Minutes Lost (CML). Figure 10.17 shows how the CI and CML performance has changed over the time the intervention took place.

The first thing to note is that the number of incidents on the whole has risen. This may be due to the fact that this is a lagging measure, and that the actions highlighted to improve this metric such as improving the overall performance of the network will take a number of years to realise any benefit. However, over the same time period CML have reduced, this is a positive sign, as even though the number of incidents has increased the time customers have been without power has decreased. This has a direct relationship to the capability data discussed above; as the end-to-end time for a fault decreases, it is plausible that the time customers are without electricity will also reduce.

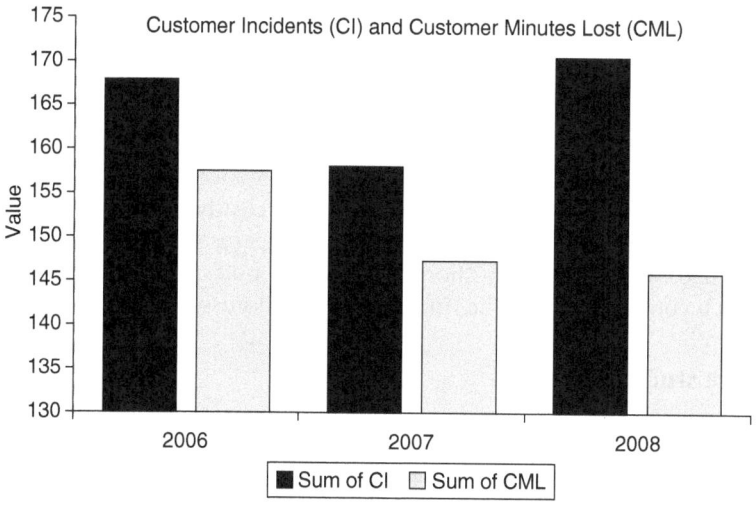

Figure 10.17   CI and CML performance

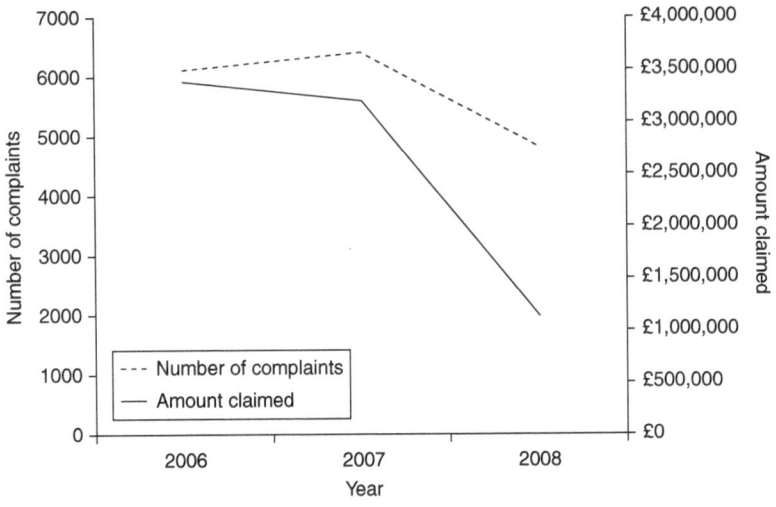

Figure 10.18   Number of customer complaints and amount claimed

Customer complaints

The number of customer complaints is important to the DNO, even though it is a lagging measure. If customers are dissatisfied with the DNO, it is reasonable to assume they are unlikely to buy products and services from the parent company and associated business units, a concept discussed by Zeithaml et al. (2006) in terms of how improving customer satisfaction leads to improved relationships with customers, leading to greater customer loyalty and an increase in a company's competitive advantage. Figure 10.18 shows that over the period of the intervention, the number of customer complaints reduced significantly.

The amount claimed by customers is also clearly important as this directly impacts the profitability of the organisation. The amount claimed fell by over £3 million during the period of the intervention, which counted towards the financial saving discussed above.

Case study discussion

Effect on stakeholders

Clearly, the use of Seddon's lean systems methodology has delivered some positive results for the R&R element of the DNO. The capability measures VI, VO, and TET have all seen improvements, and as these measures were derived from the purpose which was defined by studying customer demand, there should be a positive impact on the customer. Seddon's methodology concentrates on a single purpose, that from the customers perspective, which Jackson (2005) suggests is one of the weaknesses of the methodology, as other stakeholders are not considered.

In this particular case study, although other stakeholders have not been considered, by focusing on the customer, performance relating to them has improved. Taking the senior management of the DNO and shareholders as an example, one of their primary concerns is EBIT (Earnings Before Interest and Tax), which is demonstrated in the company metrics and reinforced at team briefs, etc. This intervention has delivered considerable net savings, which will have a positive impact on the EBIT of the DNO, therefore although they were not included in this intervention; it is fair to assume that these stakeholders would see the intervention favourably. A similar assumption can be made about the industry regulator, OFGEM, as performance against their CML measure has improved, and improvement in CI is likely to be seen as the effect of improvements to the network flows through to the metrics.

One of the advantages of Checkland's SSM is its pluralist approach to participants as demonstrated in Figure 10.4. Checkland uses the pneumonic CATWOE to define a 'well-formed root definition' (1999, p. 225), therefore to test how effective Seddon's approach is at building a comprehensive 'root definition', the results of 'Check' will be compared against a root definition defined using CATWOE.

Table 10.1 appears to back up Jackson's suggestion that Seddon's methodology does not produce a comprehensive 'root definition' when compared to SSM, it is likely this will be the case if compared to Ackoff's interactive planning. It does appear from this case study however that it is not necessary to form such a comprehensive root definition, as the outcome from Seddon's methodology still generated a positive result for other stakeholders. This single case study is not sufficient to prove this, and a more in-depth study of a full SSM intervention compared to a full lean systems intervention on a similar system would at least be required.

Some learning points

Whilst the intervention delivered some significant improvements, there was still opportunity to have delivered more. Not all the DCs were successfully 'rolled in' in a way that the fundamental understanding of Seddon's methodology was understood. Therefore, little improvement has been seen at some DCs and due to the lack of knowledge transfer it

Table 10.1   Comparison of CATWOE and Check

CATWOE Element	CATWOE Result	Comparison to 'Check'
C: Clients or customers	• Home electricity customers • Business electricity customers • Public services (street lights, hospitals, etc.) • UK Government	Purpose defined in 'Check' is inclusive of all these customers as they all contribute to the demand studied
A: Actors, the participants in the system	• R&R craft team (DNO) • R&R management • Wider DNO management • Logistics supplier • Materials supplier • Sub contract civil supplier • Customer/general public	These actors were all involved to some degree in 'Check'
T: Transformation, what the system sets out to do	• Repair and improve the electricity network • Generate profit	The purpose defined in 'Check' does not take account the need to generate a profit
W: Weltanschung, the outlook, framework, or beliefs of the system	• We need to make a profit • We need to meet regulator targets	This is evident from the *de facto* purpose defined in 'Check'
O: Owners, who owns the system	• Shareholders • Senior management • Parent company • Government	Not all the owners are considered in 'Check', however there effects are seen in some of the system conditions
E: Environment, where does the system boundary lie	• The R&R activity at the DNO in question. The end-to-end process from receipt of fault to full resolution, including areas which have a significant effect on the system such as procurement	This element is covered by mapping the flow in 'Check'

is unlikely they will continue round the 'Check Plan Do' cycle making continuous improvement as intended. Also some of the system conditions are still in place such as the financial targets for both the R&R team and the supplier; this may indicate that management thinking was not sufficiently challenged.

Conclusions

The case study shows that although Seddon's lean systems methodology may have not been implemented fully, it has shown improvements in the capability of the system against its defined purpose, its financial performance, and its performance against OFGEM targets. If the methodology had been applied 'fully' and, for example, management thinking had been more successfully challenged, then it is conceivable that even greater improvements could have been realised. However, further research would be required to determine if this is the case.

The case study also brings into question some of the suggested limitations of Seddon's methodology, such as its focus on purpose from only one perspective; it actually shows that although only one perspective was focussed upon, that of the customer, improvements were seen in areas of interest to other stakeholders as well. However, there is insufficient evidence available to draw any absolute conclusions and further research is required.

A single case holistic case study has limitations in its ability to be generalised, and therefore additional research is required to determine the effectiveness of Seddon's lean systems methodology, and whether the limitations suggested by Jackson (2005) when compared to SSM and interactive planning are valid.

References

Aackoff, R. L., Magidson, J. and Addison, H. J. (2006) *Idealized Design: How to Dissolve Tomorrow's Crisis...today,* Upper Saddle River, N.J., Wharton School; London: Pearson Education [distributor].

Ashby, W. R. (1964) *An Introduction to Cybernetics.* Methuen.

Beer, S. (1980) *The Heart of Enterprise.* Chichester: Wiley.

Bertalanffy, L. V. (1969) *General System Theory: Foundations Development Applications.* Braziller: New York.

Bicheno, J. (2008) *The Lean Toolbox for Service Systems.* Picsie Books: Buckingham.

Bicheno, J., Catherwood, P. and James, R. (2005) *Six Sigma: and the Quality Toolbox for Service and Manufacturing.* Picsie Books: Buckingham.

Boulding, K. E. (1956) General Systems Theory – The Skeleton of Science. *Management Science,* 2, 197–208.

Checkland, P. (1999) *Systems Thinking, Systems Practice : a 30-year Retrospective,* Chichester: John Wiley.

Checkland, P. and Scholes, J. (1990) *Soft Systems Methodology in Action.* Chichester: Wiley.

Churchman, C. W. (1968) *The Systems Approach.* Delacorte Press: New York.

Davis, R. (2009) System Archetypes – Unpublished electronic media.

Flood, R. L. (1999) *Rethinking the Fifth Discipline: Learning within the Unknowable.* Routledge: London.

Hoos, I. R. (1972) *Systems Analysis in Public Policy: A Critique.* University of California Press: Berkeley.

Jackson, M. C. (2003) *Systems Thinking: Creative Holism for Managers.* John Wiley & Sons: Chichester, West Sussex

Jackson, M. C. (2005) A Systematic Approach to Service Improvement – Evaluating Systems Thinking in Housing, Appendix 1 Systems Thinking: an overview by Professor Michael Jackson. In Minister, O. O. T. D. P. (Ed. London)

Jackson, M. C., Johnston, N. and Seddon, J. (2008) 'Evaluating Systems Thinking in Housing'. *Journal of the Operational Research Society,* 86–197.

National-Grid (2008) Report to the Gas & Electricity Markets Authority GB Transmission System Performance Report 2007–2008, http://www.national-grid.com/uk/Electricity/Info/performance/

National-Grid (2009) National Grid Electricity Transmission Plc 2009 GB Seven Year Statement http://www.nationalgrid.com/uk/Electricity/SYS/

OFGEM (2008) 2007/08 Electricity Distribution Quality of Service Report. Office of the Gas and Electricity Markets.http://www.ofgem.gov.uk/

OFGEM Office of the Gas and Electricity Markets (OFGEM) website Web Page 11/06/2009 http://www.ofgem.gov.uk/Networks/ElecDist/Pages/ElecDist.aspx

Seddon, J. (2005) *Freedom from Command & Control :Rethinking Management for Lean Service,* Productivity Press: New York.

Senge, P. M. (1990) *The Fifth Discipline: The Art and Practice of the Learning Organization.* Century Business: London.

Spitzer, D. R. (2007) *Transforming Performance Measurement: Rethinking the Way we Measure and Drive Organizational Success.* AMACOM: New York; McGraw-Hill: London [distributor].

Von Bertalanffy, L. (1972) The History and Status of General Systems Theory. *Academy of Management Journal (pre-1986),* 15, 407.

Wheeler, D. J. (1993) *Understanding Variation: The Key to Managing Chaos,* SPC Press: Knoxville, Tennessee.

Womack, J. P., Jones, D. T., Roos, D. and Massachusetts Institute Of Technology. (1990) *The Machine that Changed the World.* Rawson Associates: New York.

Zeithaml, V. A., Bitner, M. J. and Gremler, D. D. (2006) *Services Marketing: Integrating Customer Focus across the Firm.* McGraw-Hill/Irwin: Boston.

11
Improving Software Project Management in Bureaucracies

Peter Middleton and Brendan O'Donovan

The purpose of this chapter is to describe an example of how current system development methods do not accurately capture an organisation's IT requirements. It is argued that current best practices in both the Agile and CMMI/planned approaches to software development routinely fail to understand the true nature of their client organisations.

The plan driven approach is based in the systems engineering and quality disciplines and was developed to reduce the number of late, over budget, and poor quality software projects. In contrast, the Agile movement emphasises communication, cooperation and seeking emergent solutions. What is original about this chapter is that it shows the traditional requirements gathering process that both approaches use, which is essentially asking users and sponsors what they need, is badly flawed.

The evidence presented is a case study of a project in a housing bureaucracy which implemented remarkable performance improvements. It did this by using the customer focused concepts of failure demand, system conditions, and total end-to-end times to gain a totally new understanding of its current capability to serve customers. These concepts provided insights which then guided significant change.

The conclusion is that by applying this variant of systems thinking it is possible to routinely produce large performance gains in bureaucracies. It is therefore necessary for both Agile and plan-driven approaches to widen their scope to take a systems rather than just a software perspective.

Introduction

To improve software projects the issue of project definition and selection must be addressed. To do this the current capability of an organisation to deliver services to its customers needs to be rigorously measured and monitored. This deeper understanding of how an organisation is currently functioning can be used to select the correct project and then develop better requirements. This approach explicitly addresses the important stakeholder and external context issues. The project management skills of scheduling, resource allocation, and control, must be seen in this broad context. The management of stakeholders is important for project success, yet this may not happen effectively with current approaches.

This case study shows how just asking for requirements is too superficial and may well lead to the wrong project being selected or to misleading requirements being captured.

Literature

Boehm and Turner (2004) formulate the current state of software development in these terms:

> In the last few years, two ostensibly conflicting approaches to software development have competed for hegemony. Agile method supporters released a manifesto that shifts the focus from traditional plan-driven, process-based methods to lighter, more adaptive paradigms. Traditional methods have reasserted the need for strong process discipline and rigorous practices. (p. xix)

The way these two methods approach the critical issues of project selection and stakeholder management is instructive. In the Agile approach, Auer and Miller (2002) state that the customer's role is to: '*determine scope, release dates and priorities*' (p. 5). Schwaber and Beedle (2002) state that: '*Product backlog content can come from anywhere: users, customers, sales, marketing, customer service, and engineering can all submit items to the backlog; however, only the Product Owner can prioritize the backlog*' (p. 7). Cockburn (2002) asserts the importance of '*Individuals and interactions over processes and tools ... customer collaboration over contract negotiation*' (p. 217).

The plan-driven approach to project selection and definition is remarkably similar. Using the Capability Maturity Model Integration (CMMI) as an example of the plan-driven approach gives the following: Chrissis et al. (2003) states:

> Requirements are the basis for design (p. 465). ... Frequently, stakeholder needs, expectations, constraints, and interfaces are poorly

identified or conflicting. ... To facilitate the required interaction, a surrogate for the end user or customer is frequently involved to represent their needs and help resolve conflicts. (p. 469)

Ahern et al (2004) in 'CMMI Distilled' states that: 'Requirements are managed and inconsistencies with project plans and work products are identified' (p. 61). The Carnegie Mellon University, Software Engineering Institute's view is more starkly put in their earlier CMMI literature: *'Analysis and allocation of the system requirements is not the responsibility of the software engineering group, but it is a prerequisite for their work.'*

The weakness in both the Agile and plan-driven approaches is that they largely rely on the experience of the customer for project selection. But as Deming (1982) pointed out: *'Experience without theory teaches nothing. In fact, experience can not even be recorded unless there is some theory, however crude, that leads to a hypothesis and a system by which to catalogue observations'* (p. 317).

Both Agile and plan-driven approaches do not seriously seek a deeper understanding of an organisation's current capability to serve its customers. Both approaches rely largely on opinion for project selection and requirements formulation. Stakeholder management is interpreted to mean to capture requirements in either quick software iterations or in detailed documentation. But there is little attempt made to assist stakeholders to understand their organisation as a system. The result of the methodologies is therefore often just to code a suboptimised status quo; expensively replicating existing poor processes in 'IT concrete'.

The requirement to align project deliverables with business strategy objectives such as time-to-market, quality, and cost is clear. The need for projects to respond as the business strategy alters to reflect environmental changes, such as market shift, is also identified (Srivannaboon and Milosevic, 2006).

The software metrics used to record progress are often just internal technical measures. These are not linked to the external environment and so do not show when a project is drifting away from strategic objectives. The need is identified to use more organisation-centric numbers (Frederiksen and Mathiassen, 2005). This view is support by the benefits identified from improving cross-functional team interaction by using common metrics, rather than each functional specialist each with their own set of metrics (Chen, 2007).

Ohno (1978) showed that attention should be on the flow of work through an organisation with a constant focus on removing any step that did not add value to the customer. Deming (1982) demonstrated the advantage of managing organisations as systems. He showed the dangers of focusing on functional departments with their own performance targets as chronic suboptimisation of the overall system would result. It is paradoxical that much systems analysis does not capture the functioning of the system at all. It focuses on the needs expressed by people in organisational silos. The projects are therefore often doomed before they start.

Seddon (2003, 2008) built on Ohno's and Deming's work to develop a method for implementing 'systems thinking' in service organisations. The first step was to ensure the organisation was understood as system. The perspective for this understanding must be the customers, rather than that of the existing organisational hierarchy. While this can be challenging, in the long term without customers the organisation will cease to exist. This is quite different and more objective than asking people what they need.

Once the organisation is understood as a system then many of the previous problems that apparently required software projects may well have been 'dissolved' (Ackoff, 1999) and simply no longer exist. The project selection process can now start from a solid foundation. When the organisation is understood as a system the improvement effort can be targeted to where it has most leverage, rather than to the functional department with the most political clout.

Finally, the changes can be implemented and tracked against the initial capability analysis of the system. This allows effective and objective monitoring of results, so directing any further refinements required. The systems thinking approach is to use time-series measures of capability to show a system's predictable capability to serve its customers. By focusing on the reason for and the pattern of customer interactions, significant waste caused by faults in the system itself is usually identified.

Research methodology

To examine the benefit of first analysing a bureaucracy as a system from a customer's perspective a case study was undertaken. While the literature (Deming, 1982; Ohno, 1978; Seddon, 2003, 2008) was convincing, the need was to see if this version of systems thinking could be incorporated into software development practice. A UK organisation, Vanguard Consulting Ltd, claim to be able to transform an organisation's performance by the application of systems thinking. One of their projects was studied in detail.

The researcher joined the project team to shadow an experienced consultant. He worked as an embedded member full time for approximately six weeks between 29 October 2007 and 9 January 2008. He observed the work of the consultant, project team, Operational Managers and leaders at various stages of the intervention. Interviews were recorded for later analysis. Copies of operations data and analysis were retained to allow the change in performance to be independently monitored. He had unrestricted access to all project data. He stayed in the same hotel as the consultant during the week to learn more about the assignment. He wrote up notes and collected documentation on a daily basis. He then kept in contact with the organisation for over eight months after completion to monitor results.

The project selected was a housing association that provides homes and services to more than 25,000 people. It works in 16 local authority areas and manages over 9,000 properties. It has over 300 staff providing services to their residents and other customers. The housing repairs work to be analysed was concentrated in the South Wales area of the UK.

Client staff had attended Vanguard's introductory training and a relationship had developed over the previous 18 months. A consultant had been to visit the client in September 2007 in order to complete two days scoping work. The explicit aim of the organisations' directors was to improve service to their customers, their residents.

Case study

The approach taken was to start by looking at the purpose of the system from the customers' point of view. It is vital that client staff and managers carry out this analysis themselves in order for them to understand the significance of the current design of the system and to believe the results it caused. They are used to working in a top down, target driven hierarchy and so parts of the system will have been invisible to them. This is also likely to be the first time that they will have seen detailed and objective data on the service their organisation actually delivers to its customers as opposed to performance against arbitrary standards.

The process the organisation originally used to select their IT systems was the following:

1. Gather requirements from all different departments
2. Evaluate and select requirements gathered

3. Produce a requirements document with a ranked list of requirements
4. Look at major software packages available on the market
5. Go out to tender

This is standard practice and appears reasonable, but this chapter argues that this method to determining requirements is flawed. As Deming points out, focusing on functional departments with their own performance targets leads to chronic suboptimisation. Therefore using the departments' needs as the foundation for an IT system is unwise. A stronger method is described below and was later used by the organisation to achieve significant performance gains.

Seddon's method shown in Figure 11.1 requires managers to start with looking at the purpose of the system from the customers' point of view. This is crucial for the success of the system. From this perspective, it is possible to identify 'failure demand' (Seddon, 2003). Failure demand is 'demand caused by a failure in the system to do something or do something right for the customer' (Seddon, 2003, p. 26). Examples include not doing what the customer expected, such as failure to call back, failure to turn up, or failure to send something that is anticipated; or a failure to do something right, such as not solving a problem, or sending out a form that a customer has difficulties understanding.

In order to study the existing capability of the organisation, a team of frontline workers was assembled. The team consisted of workers from across the organisation (a call centre operative, the asset management officer [who undertakes internal inspection of work and oversees planned maintenance works within a certain area] and a performance

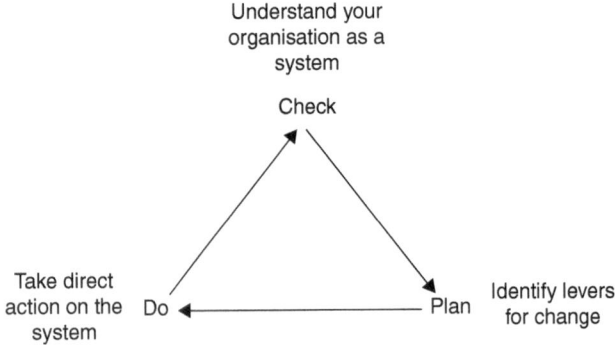

Figure 11.1   Check-Plan-Do method

management officer) and two of its tradesmen subcontractors. In parallel, both similarly sized teams of operational managers and the organisation's leaders were set tasks, meeting weekly to discuss the progress of the intervention and to report back on what they themselves had learnt.

In order to take an outside-in, customers' perspective view of their organisation, the team was required to listen to demands that came into the call centre.

1. Purpose and what matters

An initial afternoon listening to demands in the call centre gave the team an insight into the customer's housing requirements from the system. It was then that the team agreed upon what they thought was the purpose of the system: 'fix it', with the 'it' being 'my housing repair problem'. The team also looked at what matters from the customer's perspective. This was agreed to be:

- Do it quickly
- On one visit
- Do it when it suits me
- It stays fixed

With services, where the customer is involved in delivery, a standard service is not the objective. To maximise customer satisfaction and reduce cost, the way is to be organized to deliver precisely what each individual wants. Each customer will have a separate set of circumstances and constraints. This is Taguchi's (1993) concept of Nominal Value. Instead of working towards targets within tolerances, it is better to aim at perfection and reduce variation. Paradoxically this results in cost reduction.

2. Type and Frequency of Demand

The team then went back to listening to the variety of demands that came into the call centre, attempting to document the many types of calls that the centre had to handle. After around one week of this activity, where all demands were written onto flipcharts and displayed in the boardroom, it was agreed that most types of demands had been recorded, and a new tick sheet was created for the call centre operatives to use so that both the type and frequency of calls into the centre could be understood and evidenced. These tick sheets were left with the operatives to complete, and the results were steadily fed into

(%) Value demand	(%) Failure demand
12.7 - Please fix my central heating/hot water/boiler/radiator	6.1 - Can I have an order? (Tradesman)
4.9 - Please fix my door/ door furniture	5 - When will the tradesman be here?
4 - Please fix my leak/burst/overflow	3.3 - Recall – heating and plumbing
3.6 - Please fix internal and external lights	2.8 - I've reported a problem, I haven't heard anything
3 - Please fix my windows	2.8 - No access
2.9 - Please fix my taps	2.3 - I want to cancel a job
2.8 - Please fix my toilet	1.8 - Someone was supposed to come but didn't
2.2 - Please fix my electrics	
1.6 - Please fix my storage heaters/ electric fires	1.5 - Please extend the target time. I have not been able to get in
1.5 - Please fix my smoke detector	1.5 - Please explain (contractor)
	1.3 - Recall – carpentry and wet trade

Figure 11.2   Top value and failure demands in the system

spreadsheets in order to allow the team to analyse the results at the end of the Check period. It was here that the Value and Failure Demand shown in Figure 11.2 were separated.

Alongside this breakdown of type and frequency of demand, there was a geographical analysis of demand; mapping the number of repairs per property in various estates. This was to help understand whether demand is predictable or unpredictable. Taken altogether, this analysis provided some of the knowledge of the system required to allow services to be designed against demand.

When these results were analysed it showed a 37 percent failure demand in the system. Failure demand is 'demand caused by a failure to do something or do something right for the customer' (Seddon, 2003 p. 26). Failure demand is complete waste for both the organisation and the customer. The reduction or elimination of failure demand will therefore rapidly reduce costs and improve service.

3. Capability

Figures were sought from the current system that could show the volume of jobs requested day-by-day for the contractors to repair over a 12-month period. This was then mapped as time-series data in the form of statistical process charts (SPC) or Capability Charts as shown in Figure 11.3.

The chart in Figure 11.3 shows that jobs requested fell sharply over Christmas with a corresponding spike in early January. This pattern would be predictable every year. If this spike is ignored it can be seen

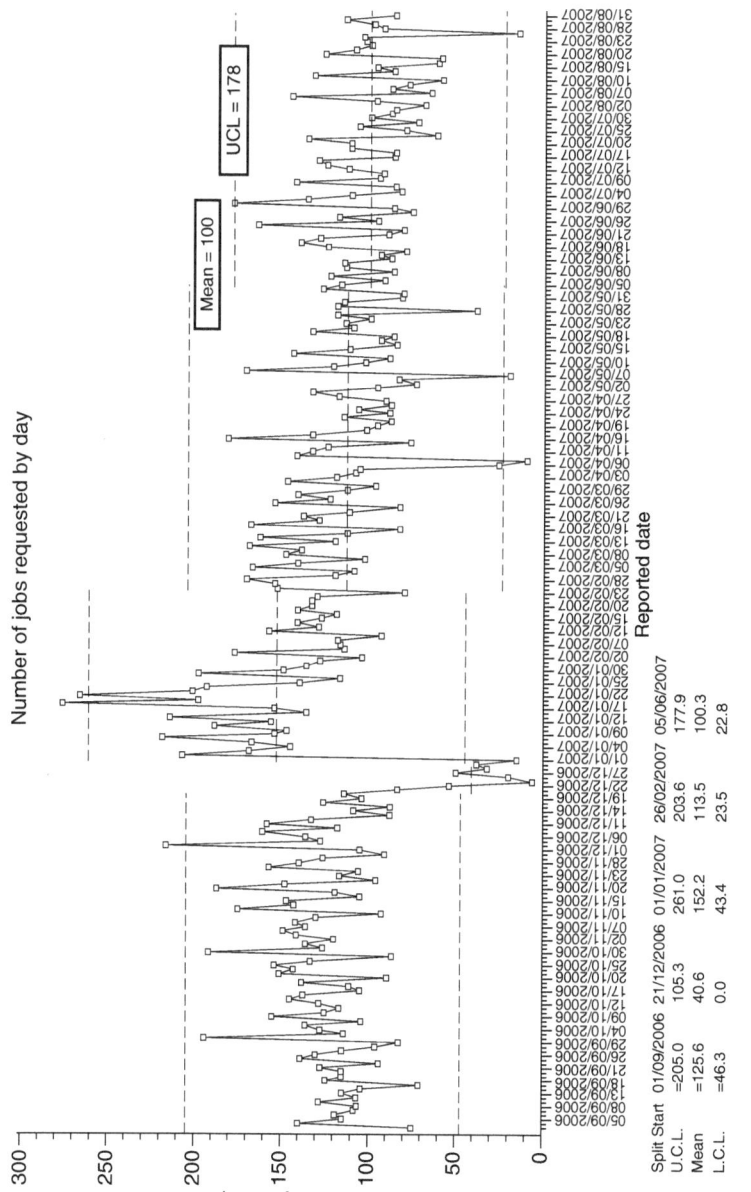

Figure 11.3 Capability chart of housing repair jobs requested by day (Sept 06–Aug 07)

that daily demand has a mean of between 100 and 125 jobs requested per day. The variations would be due to the seasons, weather, public holidays, sporting events, measurements taken and random noise in the system. It would be normal to expect a spike of demand on Mondays, after the office was closed for the weekend. We would expect repair requests to be lower in the summer when the weather is better and people are away on holiday. A precise understanding of demand is essential to be able to design a system to handle the demand.

A capability chart shown in Figure 11.4 of jobs completed per day for the same period from September 2006 to August 2007 was also made. Comparing Figures 11.3 and 11.4 it is evident that the inputs and outputs into the system are quite similar. Comparing the mean of the number of jobs coming into the system with the mean of the jobs completed, shows the jobs completed each day are actually exceeding the new ones arriving. Despite completed jobs being greater than the new jobs arriving, there was always a significant pool of work held in the system. The surplus of completed jobs over reported jobs was due to a job sometimes being reclassified into two or more jobs. The important thing to note is that the system had the capability to deal with the work coming into the system; yet there was always a substantial backlog. Figure 11.5 allows us to obtain real insight into the precise level of service the customers were experiencing.

To analyse the actual customer experience the capability chart in Figure 11.5 was constructed by pulling out a sample of jobs from the same time period. This sample traced the date from when a customer first made contact with the organisation to the date when the job was successfully completed. This customers' eye-view was presented as a capability chart showing actual end-to-end times they experienced. The results indicate that far from beating the Welsh Assembly Government's target times of 28-day maximum, the mean time is 50 days with an upper control limit of 122 days. If a guarantee was to be offered to residents when their repair would be fixed, this system could only guarantee that it would normally be reliably and predictably fixed within 122 days.

4. Flow

To understand why the jobs were taking so long to complete, especially when more jobs were recorded as completed each day than entered the system, the flow of work was mapped on flipcharts across a wall. Ohno's (1978) experience was that to understand a system the flow of work

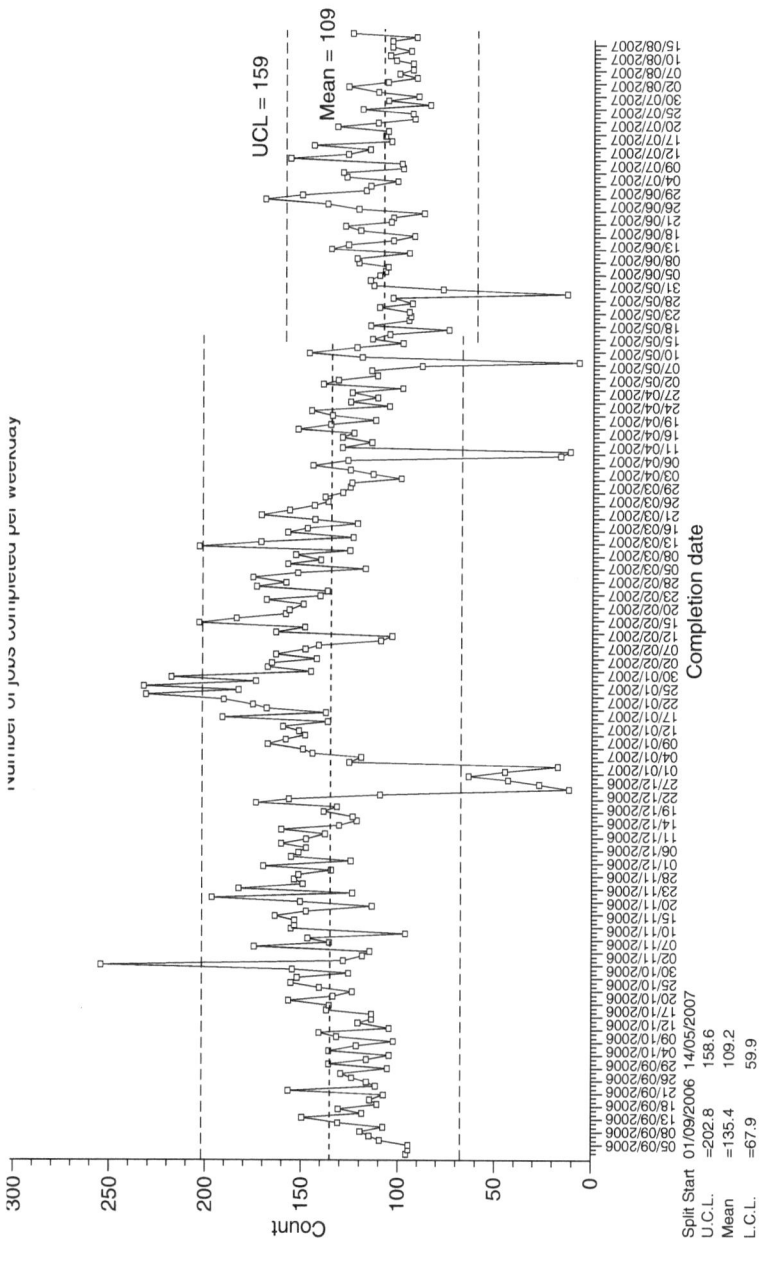

Figure 11.4 Capability chart of jobs completed per day (Sept 06–Aug 07)

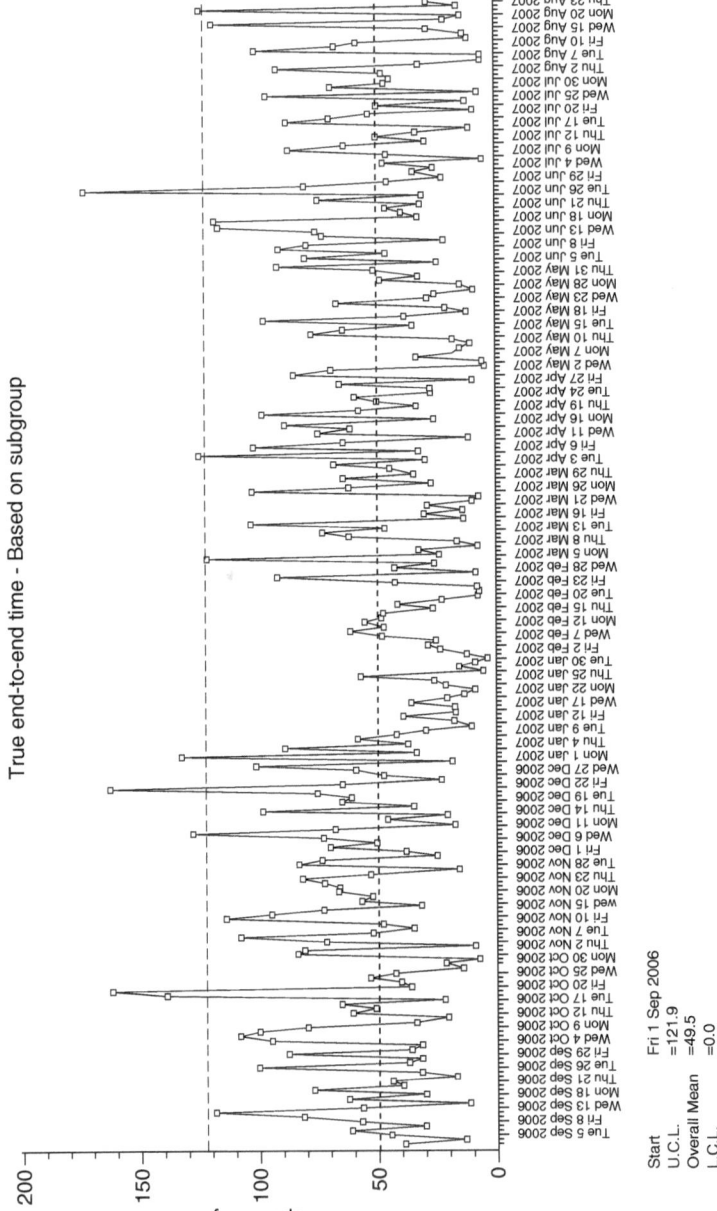

Figure 11.5 End-to-End capability from the customers' perspective (Sept 06–Aug 07)

must be followed through the system. In this case the flow was from the call centre, out to the subcontractor to make the repair, through a visit by an inspector to ensure the work was satisfactory, to claims for payments for materials and labour and finally to the payment being made. The process went through 38 steps. The Check team also gathered further information about rework in the flow (e.g. how often work had to be returned due to incorrect information). Of these 38 steps, only 3 could be seen as providing value to the customer:

• diagnose problem
• gain access to property
• fix the problem

There were 17 handoffs (one person passing work onto someone else) in this process. The team also analysed the payment process between the housing association and its subcontractors. Isolated as part of the process, this took another 26 steps and 12 handoffs to make the payment, with the range of times taken for payment being 18 to 260 days. This was all waste from the customer's perspective. The analysis of flow showed where all the time and money was going in the system. It explained why with ample resources available that the average time to successfully fix a repair was 50 days with variance giving an upper control limit (UCL) of 122 days (four months). It showed a system that was wasteful and with wide variations in performance.

5. System Conditions

The reason the system was performing so poorly was partly due to the system conditions it was operating under. System conditions are the things that explain why a system behaves in the way it does. From the analysis of flow the system conditions were identified as:

• the method of paying contractors (full of inspection and double checking)
• targets (such as the current measures, including the Welsh Assembly Government's requirement to break the work down into 1 day, 7 days, 21 days, and 28 days)
• the functional specialisation of the workers
• IT systems

The operational managers examined their current measures, and came up with their most important three.

The current top 3 most important measures as judged by managers	1. Percent of repairs completed on time 2. Customer satisfaction 3. Percent urgency codes (1/7/21/28 days)

The managers were to consider the performance measures that they currently used, and then analyse the impact that they have on the system. The managers had to consider:

- What did these measures tell them?
- What decisions did they make based on these current measures?
- Were they appropriate from a customer perspective?

By using the new information gathered under 'what matters' to customers, they came up with these new measures:

What matters to the customer?	New Measures
Do it quickly	End-to-end time
On one visit	Percent of first visit fix
Do it when it suits me	Percent do it when I want you
It stays fixed	Percent it stays fixed

It was vital that the measures used to manage the organisation were changed as they are 'system conditions' that drive how the system performs. They are not neutral. These new measures were to be used during redesign. Figures were gathered to see what current performance was against these measures, and this was shown in a capability chart. It is important to note that systems conditions are not mentioned in either the Agile or plan-driven literature. These methodologies therefore routinely miss an essential component that drives system performance.

The literature review contains direct quotes of current best practice on how information systems should be specified. For example, it is the customer's role to 'determine scope'. The problem with this approach that we wish to demonstrate is that often the client sits within one component of the overall system and so their view often does not reflect the impact of the total system on the customer. Their requirements therefore reflect their position within the current organisational structure, which is often suboptimised and dysfunctional. Unless this is explicitly checked for, by analysing how customers are actually being served, *before* requirements are collected, the requirements will often be misleading and damaging.

6. Leaders' Review

All of the information gathered was presented jointly by the operational managers and the Check team to the leaders, who have to decide whether their system has room for improvement, before any attempt is made at redesign. The result was a powerful expose of the suboptimal performance of the system delivered by those who work in it. The leaders, in this case, were ready to agree that their system was flawed, and that they wished to try and improve it. This allowed the organisation to move into 'redesign'.

7. Redesign

Once the leaders had given their assent to starting the redesign process, the team began looking at how to do just the value work, with the permission to experiment with new methods of achieving this. The purpose of redesign is 'to learn the art of the possible in pursuit of perfection'. Although they started with a blank slate, the team followed a few principles and rules for the redesign. These were:

Principles for redesign	Rules
Only do the value work	Don't break the law
Customer defines what matters	Don't change the IT until it's fully understood how the system works and impacts the customer.
Single piece flow	
Work as one	
Minimise hand-offs	

The team set up a room where they could experiment with the method of doing the work. They had calls directed from the call centre that were from clients that lived within the Cardiff area. This gave the team a manageable number of calls to work and experiment with so they could learn how to improve the service. The team responded to whatever demand came in to them. They would take the demand, ask the caller when they wanted the job done, plus whether there was anything else that could be done at the same time, and then one of the tradesmen from the team was dispatched to complete the task. It was this simple initially, meaning that the measures showed massive improvement from those found during Check. As the demands increased, it became necessary to think about the way that appointments were made by the contractors, and what geographical areas would be covered by which contractor. In order to collect information to improve the system, they would follow up the work with a call to the resident, asking them:

- to score the level of service they received on a scale of one to ten
- ask whether the repair had stayed fixed
- ask if the work was done when it suited them
- ask if anything could have been changed to help them more

Also, the end-to-end time was recorded, and the tradesman reported whether it had been fixed on the first visit.

The Operational Managers were also involved with the work, taking time to find out about the new methods that the redesign team were using, and considering how they may be scaled up for use by the rest of the organisation. They were also looking at demand to see how much of it was predictable. They also solved problems that were encountered by the team.

In turn, the leaders are managing by walking around, learning more about the work and providing a steer to the Operational Managers on problem solving. For example, the Chief Executive gave her support to experimenting with a new recharging procedure.

The early results achieved by the redesign team over the first 70 jobs were spectacular. End-to-end times were now averaging 1 day, with the upper control limit being 3 days. When compared with the mean of 50 days and upper control limit of 122 days, massive improvement can be clearly seen. Customer satisfaction is averaging around 90 percent.

After 8 months of the redesign, in September 2008 the following figures were recorded:

- Costs falling by £20 (US $30) per job from previous system
- Total annual cost savings of £600,000 (US $875,000) for the organisation
- Customer feedback average score of 90 percent
- Customer average end-to-end times fell from 50 to 6 days

The important point is that the concept of achieving an understanding of demand and the capability of the current system to handle it has shown how significant changes can emerge. It is far stronger than just asking for requirements from people with restricted insight into their own organisational system. These results were achieved with no change to the IT systems.

Discussion

The following work was carried out:

- Six weeks for the team to understand customer demand, work flow and organisational capability
- One day of presenting these findings and the organisation deciding if it wanted to proceed to improve their system
- Experiment with work redesign and 'roll in' to the rest of the organisation

It must be stressed that there are no distinct 'as-is' and 'to-be' positions, as the learning is on-going and emergent. Even though the consultant had great experience of similar systems, it is recognised that each system is distinctly different and therefore imposing ready-made solutions would be a recipe for failure. This is both from a systems theory position of wanting to optimise the performance of a system (which would be impossible without understanding the unique) and also from an interventionist perspective, as it would mean that the client was not equipped to understand their own system and therefore unable to continue to improve. The change would be less likely to be sustainable.

Starting from a customers' perspective on how the whole system functioned, it produced annual savings of £600,000 and an eight-fold improvement in performance, with no expenditure on IT. There are still unresolved issues, but by continuing to apply a systems thinking approach, future decisions can be made based on objective knowledge of their capability. This work supports the other accounts of performance transformations due to applying systems thinking (Deming, 1982; Seddon, 2003, 2008).

Conclusion

What is significant about this case study is that when the organisation initially used the conventional method to obtain requirements, the resulting systems performed poorly. This approach is essentially to ask the various departments what they need, compile these into a requirements document, and then go out to tender for a system.

When the alternative method described here, of collecting precise data on how the total organisation was performing as a system from a customer perspective, then great performance gains were reported. Software developers are too narrowly focused on looking primarily to deliver software, rather than better systems. They need to first establish a deep understanding of the pattern of demand the system is to handle and determine the current capability of the organisation to handle it. From this basis the overall system can be redesigned, and then software specified if necessary. The deep problem this work illustrates is the one identified by Boehm and Turner (2004):

> Unfortunately, software engineering is still struggling with a 'separation of concerns' legacy that contends that translating requirements into code is so hard that it must be accomplished in isolation from people concerns. In today's and tomorrow's world, where software decisions increasingly drive system outcomes, this separation is increasingly harmful. (p. 153)

The way forward therefore for both Agile and plan-driven methods is to start with the version of systems thinking described here. Both approaches should recommend that software developers work first to truly understand demand, and second establish the current capability of the bureaucracy to provide services to its customers. After generating this new knowledge the needs and IT requirements of an organisation may well be redefined. From this solid foundation the necessary systems can be designed and implemented in collaboration with the organisation.

References

Ackoff, R. L. (1999) *Re-Creating the Corporation: A Design for Organisations in the 21st Century*. Oxford University Press, New York.

Ahern, D. M., Clouse, A. and Turner, R. (2004) *CMMI Distilled: A Practical Introduction to Integrated Process Improvement*. Addison-Wesley: Boston.

Auer, K. and Miller, R. (2002) *Extreme Programming Applied*. Addison-Wesley: Boston.

Boehm, B. and Turner, R. (2004) *Balancing Agility and Discipline*. Addison-Wesley: Boston.

Chen, C. J. (2007) 'Information Technology, Organisational Structure, and New Product Development – The Mediating Effect of Cross-Functional Team Interaction.' *IEEE Trans. Eng. Mgt.*, 54(4), 687–696.

Chrissis, M. B., Konrad, M. and Shrum, S. (2003) *CMMI: Guidelines for Process Integration and Product Improvement*. Addison-Wesley: Boston.

Cockburn, A. (2002) *Agile Software Development*. Addison-Wesley: Boston.

Deming, W. E. (1982) *Out of the Crisis*. Cambridge: MIT Press.

Frederiksen, H. D. and Mathiassen, L. (2005) 'Information-centric Assessment of Software Metrics Practices.' *IEEE Trans. Eng. Mgt.*, 52(3), 350–362.

Ohno, T. (1978) *Toyota Production System*. Productivity Press: Portland, Oregon.

Schwaber, K. and Beedle, M. (2002) *Agile Software Development with Scrum*. Prentice Hall, New Jersey.

Seddon, J. (2003) *Freedom from Command and Control*. Vanguard Press: Buckingham.

Seddon, J. (2008) *Systems Thinking in the Public Sector*. Triarchy Press: Axminster.

Srivannaboon, S. and Milosevic, D. Z. (2006) 'A Theoretical Framework for Aligning Project Management with Business Strategy.' *Intl J. of Project Management*, 24(6), 493–505.

Taguchi, G. (1993) *Taguchi on Robust Technology Development: Bringing Quality Engineering Upstream*. ASME Press: New York.

12

The System is Always Greater than its Parts: Using the Theory of Constraints and Factory Physics to Transform Manufacturing Operations

Justin Watts

This chapter is concerned with seeing the systemic implications of a transformation programme in the manufacturing environment of a large multinational paper-based packaging supplier. It is argued that understanding variability was a key success factor in making the necessary improvements within the organisation. Once this systemic perspective had been taken, then the ideas of Factory Physics and Theory of Constraints were applied to the work. Reduced cycle time and increased throughput was created by changing policies and physical factors, managing the constraint with a minimum buffer to protect throughput, and using a maximum buffer to reduce arrival variation. By identifying and diagnosing the greatest sources of variation in a production system and understanding their impact on flow, it is shown how transformation can be taken to a level unobtainable purely through the application of tools.

Introduction

This chapter shows that concentrating on improving the performance of a factory as a system is more profitable than optimizing the various constituent parts of an operation through, for example, the use of conventional 'lean manufacturing' tools. Specifically, the application of the ideas of Factory Physics (Hopp and Spearman, 2000; Mackle, 2007; Bicheno, 2006) and Theory of Constraints (Goldratt, 1990) to a large, multinational paper-based packaging supplier are explored here, and it is shown that being an exemplar of lean techniques is not in itself a route to successful transformation. Conventional, reductionist approaches to the design and management of work concentrate on optimising the performance of the separate individual functions of an organisation, often without reference to one another. In contrast, systems thinking adopts a fundamentally different analysis, focusing on the inter-relationship between the various parts of a system. This systems approach applies just as equally to manufacturing as it does for services, as exemplified by Taiichi Ohno's Toyota Production System (Ohno, 1988). This study shows how understanding variability is a key factor in the success of a transformation programme.

Literature review

The theories of Eli Goldratt (Theory of Constraints) and Hopp and Spearman (*Factory Physics*) were developed by physicists who derived their theories from first principles (Bicheno and Holweg, 2009). Goldratt's ideas found fame through his book *The Goal* (Goldratt and Cox, 1984) which told the tale of a manager in a metalworking plant who learns to use the Theory of Constraints (TOC) as a model for managing his plant as a system, ordered around identifying and ordering the constraints or bottlenecks in a process. Whereas Goldratt's ideas are written in an accessible style, non-mathematicians often have difficulty with Hopp and Spearman's *Factory Physics*. Those who decide to persevere and try to absorb the fundamental principles behind what is being taught have much to gain. Kate Mackle of Thinkflow Ltd teaches that 'Toyota did not start with the tools; they did not start with their system. They started with an unremitting focus on how to use their resources to produce as close as possible to what the customer wants to buy now; how to align the flow of production as close as possible to the flow of cash into the business' (Bicheno, 2006, p. 4) Mackle's belief is that conventional lean improvement approaches lead to:

- Local efforts tackling isolated parts of the system to no overall beneficial effect in service or profit
- A poor understanding of demand and capacity; with plants not being run to maximise throughput of bottleneck equipment
- Naïve application of tools and techniques; often in the wrong place at the wrong time
- Inappropriate delegation of leadership for improvement (Mackle, 2007)

The *Factory Physics* principles, along with the basics of Goldratt's work, the ideas of systems thinking, and the use of Mackle's concept of creating flow were all influential on the work undertaken as part of this research. In addition, the findings of Seddon (2005) and his systems thinking methods helped to give an explanation to many of the phenomena observed, such as the relationship between organizational thinking, the management of a system and its performance, or the need to aim for the creation of economies of flow rather than economies of scale.

Research method

The work documented here was undertaken within the paper and packaging industry whilst researching for a Masters degree in lean operations. The work forms a comparative case study (Yin, 2009) between two separate factories within the company.

Approach to improvement

Within this company (a large multinational paper based packaging supplier), there was a pre-existing, 'one size fits all' transformation model. The main thrust of improvement activity was aimed at increasing throughput, specifically with the aim of increasing production capacity. In other words, the aim was to sell (or at least to make) more, without any real consideration of the end-to-end system or improving flow. After time, the managers became puzzled as to what was going wrong: they appeared to be succeeding in creating capacity and yet they were not seeing the benefits to the bottom line. This piqued the author's interest.

In common with many manufacturing managers, the first approach was to reach for the toolbox of ready-made solutions. The starting point was often to diagnose Overall Equipment Effectiveness (OEE) and look to reduce the losses. OEE scores of 20 percent were not uncommon, and the range was quite large from factory to factory. However, there was no real consideration of demand. In other words, the focus was on improving OEE on machines in general to increase capacity, which in turn was driven by the need to meet the budget. Typical lean manufacturing tools and techniques such as Total Productive Maintenance (TPM), 5S and changeover reduction programs were all attempted. This did not increase capacity to the levels envisaged by the incumbent consultants who were supporting these transformations. However, at the same time some work streams (projects) were started as a result of downtime analysis that showed some of the greatest sources of lost time were actually system based: for example, waiting for material from a standard shared resource in the paper industry (normally the corrugator), a queue of material causing lost capacity after the converters (the bottleneck) due to breakdowns on a finished goods line, and the flow of material was seen as a major problem. In fact, it was the system capability which was the real problem.

Figure 12.1 shows the relationship between corrugators and converters in relation to typical paper conversion factory flows.

Figure 12.2 shows this lost time in minutes for no material week-by-week, from the supplying process (conversion down because of no material) and lost time for blocked lanes (queue) after the converter.

The use of the tools was eventually relinquished. Instead Goldratt's principles (namely Drum-Buffer-Rope [DBR] techniques) were then used to obtain a step change in improvement. In the majority of cases, on average, the corrugators (located upstream) and finished goods lines had overcapacity in terms of volume of material supplied and taken away, in comparison to the combined capacity of the converters they served. This was due to machine capability especially at the corrugators which are traditional mass production volume driven machines. It was unclear why there was so much wasted time in the rest of the value stream. It was discovered that, on the majority of occasions, the corrugators were measured and operated for their own efficiency with little consideration for the constraints in front of them. This was a major finding for the organization.

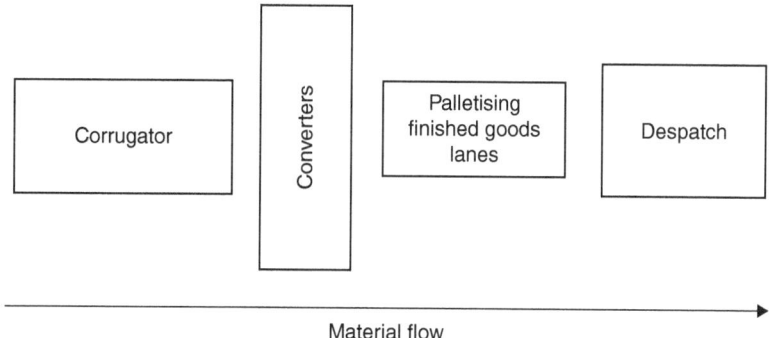

Figure 12.1 Typical paper conversion factory layouts

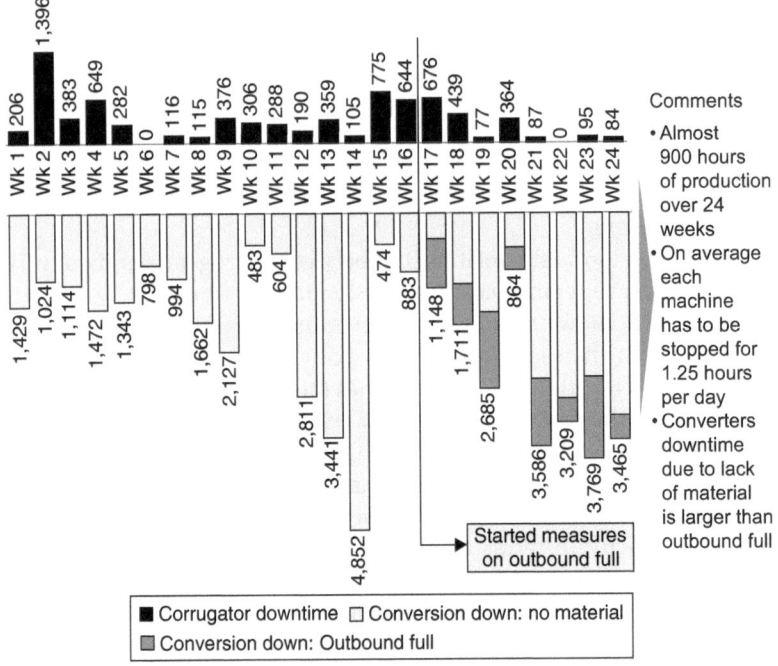

Figure 12.2   Lost time in the process

It was then realized that any improvement in OEE made through down-time reduction at the converter would in fact be dwarfed by the improvement that could be made in terms of flow and making improvements to the whole system. It would seem obvious to conventional, non-system thinkers to use tools such as TPM to increase capacity at the converters (the critically constrained resource, CCR, Goldratt and Fox, 1986, pp. 90–134) but frequently the greatest source of lost time was due to poor flow and the capability of the total system to generate throughput.

It was clear, as Mackle describes, that 'Policy and Physical Factors' (Mackle, no date, p. 37) (included in these were measures) around the running of the corrugators were a detriment to the whole system. The plants were being run in a classic MRP (Material Requirements Planning) fashion with the most expensive piece of equipment pushing the process from the back, regardless of what the constraint was doing, in the belief that greater output from the corrugators would mean greater output from the system.

A major step in improving the business performance was therefore to change the policy and physical factors determining plant use. First, it was necessary to understand that our conversion capacity determined the throughput (output) of the plants; not the capacity of the corrugator. Second, by using TOC it was possible to show that an hour lost at the converters was an hour lost for the whole system. This was essential for educating those people who determined the Physical and Policy factors. To improve this situation corrugators had to be run differently i.e. to respond to the constraint in front of it as 'synchronous manufacturing' (Goldratt and Fox, 1986, pp. 70–139). The improvements were made in the following ways:

- Changing corrugator budgeted measures, i.e. removing volume per hour measures and introducing measures for waiting for material (time) at the converters (the constrained resource) as they determined throughput.
- Understand actual machine limits at the corrugator, not what had previously been chosen as its capability to meet the budget.
- Using the MRP system in a 'leaner' way, i.e. reduce the planning horizon from a 24-hour plan to 4 blocks of 6 hours, moving away from 'economies of scale' to 'economies of flow' (Seddon, 2005, p. 20)
- Planning in a way (within the 6 hour blocks) that ensured a minimum buffer of work behind the constraint (converters)
- Applying TPM to corrugators to increase OEE and head room over the constraint along with changeover techniques to improve EPE (Every Part Every interval, a measure) and ensure the extra capacity was used for flexibility and maximize the systems capability, 'using the tools for our problems'.
- Applying TPM to finished goods lines to ensure that product was never held up once it had left the constraint.

It is important to notice that the first four changes had nothing to do with traditional 'lean' activities or tools but instead were mostly to do with changing the choices made in the business. We were moving towards using tools to improve flow and alleviate all of the associated problems encountered with lost throughput (output) due to starvation and queuing. We were using the drum buffer rope (DBR) method to schedule the plants, albeit in the first instance without a maximum WIP (Work in Progress) as the main concern was to protect throughput. This had a significant effect on plant indicators that were being measured in the organization's balanced scorecard, typically throughput (output) and OTIF (On Time in Full) (see Figure 12.3).

Improved OTIF

Stage 2 shows the change to the TOC techniques of drum buffer rope scheduling (as opposed to the lean techniques of 5S, value stream mapping and kaizen events) and the associated changes in increased OTIF and less variability. At this point the 'lean team', although not clear on the specifics, realised that there were two levels of lean transformation for our business:

- Where there was a basic level of stability that enabled identification of a constraint, an approach was needed that tackled policy and physical factors to improve flow and improve the whole system, i.e. using Drum Buffer Rope scheduling to schedule the plants to increase throughput.
- Where basic stability was not in place, the 'lean tools' would be needed and would be relevant to create stability.

As the author moved through the transformation process, it became clear that some plants were on a higher operational level than others with different levels of variability. However, diagnosing variability and really understanding its impact was not in the conventional 'blueprint'

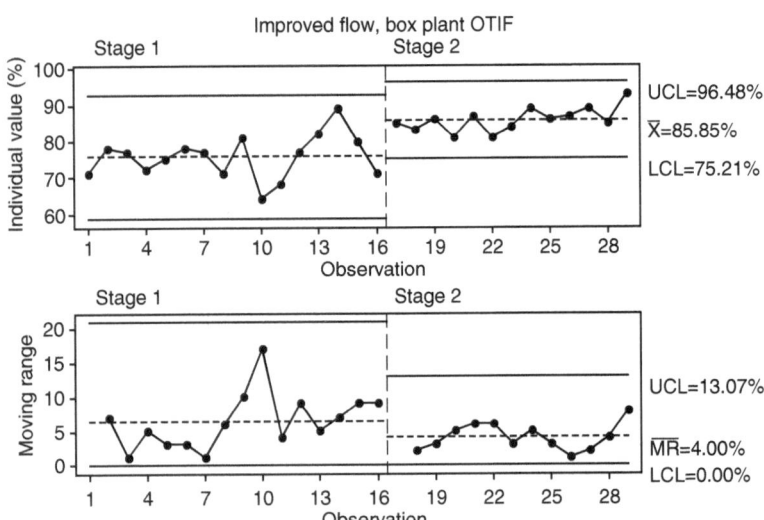

Figure 12.3 Improved flow, box plant on time in full

for transformation. How one judged whether there was basic stability was a matter of opinion. It was necessary to change the way we had been thinking about the problem.

This two-level implementation was again challenged as a result of the MSC research which coincided with the placement of the author at one of the company's supposed 'better plants'. Much work had been undertaken at the plant which focused on maintenance effectiveness, including TPM activities and lean tools such as workplace organisation and changeover reduction.

As stated earlier, the main thrust of the lean programme was to improve throughput. However, after studying the *Factory Physics* framework it became clear that we were still missing the point. This was confirmed by understanding a simple statement from *Factory Physics*: 'a production system is optimised at maximum throughput and minimum cycle time' (Hopp and Spearman, 2000, p. 219).

The business was missing the point where, even in the plants where DBR principles were being used, higher throughput levels were being created and maintained with higher levels of WIP but with longer cycle and lead times. Lead time in any form was not being measured or used as a driver for improvement. The system's capability was only being measured in terms of output improvement, whereas a true indicator of improvement would be output and end-to-end time. The lean team realized that they were guilty at this point of thinking only about 'throughput', when the essence of lean is flow: maximum throughput and minimum cycle times.

It was a realization at this stage that with the team's current understanding of lean, it was not possible to make the transformation plant leaner, i.e. increase throughput from initial stability, improved scheduling and a better system. Instead the team realised that it was required to change its collective thinking, in order to change the system and in turn to change the performance (Seddon, 2005, p. 80). This encouraged the author to study the *Factory Physics* framework in more depth.

Variability reduces system capability

As stated as one of the key laws of *Factory Physics*, 'variability always degrades the performance of a production line' (Hopp and Spearman, 2000, p. 295) whether from lost throughput or increased cycle time. Many people (including the author), thought of variation as variability in process rates (such as breakdowns, long set ups, and changes in speed) and that highly variable processes had the greatest bearing on lost throughput (starvation) and longer manufacturing lead times (queues) in a manufacturing system. To find out more, the variability in process rates was measured in the 'transformation plant' (where the author was placed) and a 'comparison plant' on a shift by shift basis, using the coefficient of variation from the *Factory Physics* framework. The coefficient of variation is calculated by dividing the standard deviation by the mean. A simple way to think about this classification is that if the average of something is one and the standard deviation is one then the coefficient will be 100 percent, meaning that variability is quite high. This *Factory Physics* classification is shown in Table 12.1.

The process variability is shown in Figure 12.4.

Figure 12.4 shows the coefficient of variation for both the corrugator and the converter (the 924 as it was known, a constrained resource).

Table 12.1 Variability class and coefficient of variation

Variability class	Coefficient of variation	Typical situation
Low (LV)	$c < 0.75$	Process times without outages
Moderate (MV)	$0.75 \leq c < 1.33$	
High (HV)	$c \geq 1.33$	Process times with short adjustments (e.g. setups)
		Process times with long outages (e.g. failures)

Source: Hopp and Spearman, 2000, p. 252.

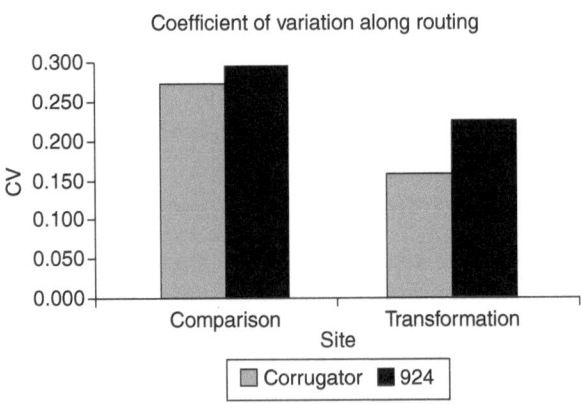

Figure 12.4 Process variability between the comparison and transformation sites

The diagram shows that the transformation plant had reduced variability by using the lean tools which had also contributed to a higher throughput rate at the converter. To diagnose the difference between the plants and what difference the lean tools were having on variability, a study was undertaken to compare the maintenance effectiveness (relating to TPM strategy) and some observations around the workplace organization of the transformation plant versus the comparison plant.

Workplace observations

The photos below show the physical differences between the two plants:

These factors contributed to a much more favourable failure pattern in terms of MTTR (Mean Time To Recovery) and MTBF (Mean Time Between Failures), changeover pattern and higher availability which gave the results for process variation above.

Failure pattern

Figure 12.5, which shows that the transformation plant has lower outage times than the comparison site.

It was clear to see that the transformation plant was using the lean tools effectively. Based on *Factory Physics* laws, the thinking of the author was that the plant with lower levels of variability and higher bottleneck rate would have better flow which would be evident in a shorter and more predictable cycle time or manufacturing lead time. As part of the research and as an extension to the current state of the transformation plant, the author then measured system time: the cycle time from corrugator to converter. This was compared to the cycle time in the comparison plant with more variability and a lower throughput rate at its slowest process (37 percent lower to be exact).

The results for cycle time (the end-to-end time) of the system are shown in Figures 12.6 and 12.7.

The table below shows that, despite the advantages achieved in terms of variability and throughput, the transformation plant had not significantly outperformed the 'non-lean' comparison plant in terms of cycle or lead times.

Plant	Rate per hour at the converter	Cycle time mean	Standard deviation
Higher variability	3601	19.64	15.75
Lower variability	5744	17.60	12.78

Comparison plant   Transformation plant

Comparison plant   Transformation plant

Figure 12.5 Transformation plant with lower outage items than the camparison site

Of these figures, it was hard to differentiate which one was using the lean tools. Apart from there being no real difference in cycle time (distances travelled were approximately the same) there was no real way of knowing how long it would take for jobs to move through the system. As for cycle time, the tools being used had done nothing to create flow through the system. These averages were of little use due to such an unstable process as the figures show.

These findings were counterintuitive. With reference to one of the laws of *Factory Physics* mentioned earlier, the transformation plant had the least variation with a more favourable failure pattern in terms of MTTR and MTBF, and a much higher throughput rate. Little's law predicts if the throughput rate is so much higher in the transformation plant and the cycle time is the same then WIP must be larger in the transformation plant. This was indeed the case (see Figure 12.8).

To back up this point an analysis of queuing time was done and showed that the transformation plant had the greatest percentage of

Outage pattern for down time attributable to breakdowns
Wk 1–20 Comp V Tran corrugator

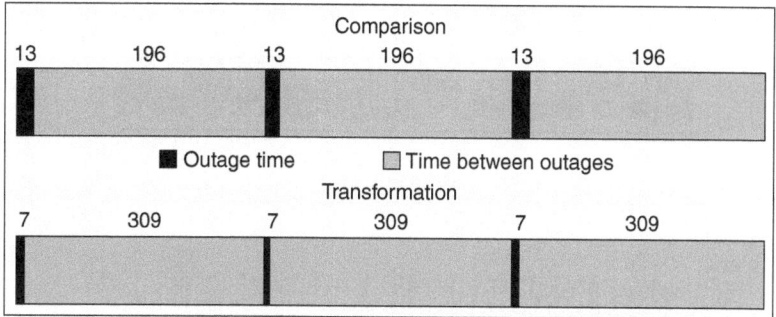

Outage pattern for down time attributable to breakdowns
Wk 1–20 Comp V Tran 924's

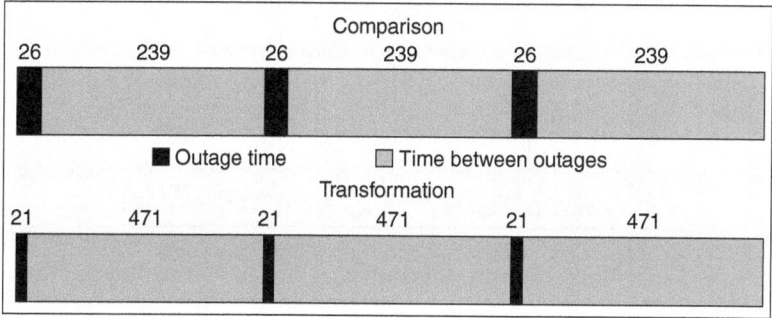

Figure 12.6   Cycle time for plant with lowest process variation

queuing time. It also showed that the majority of time was spent queuing (or waiting) in both plants.

The table below shows an analysis of cycle time percentages

	Corrugator	924	Total	Flow time	Queue time	Percent total time spent waiting
Comp	0.23	61.80	70.20	1041.60	971.40	93.26 percent
Trans	0.33	138.00	150.00	1162.80	1012.80	87.10 percent

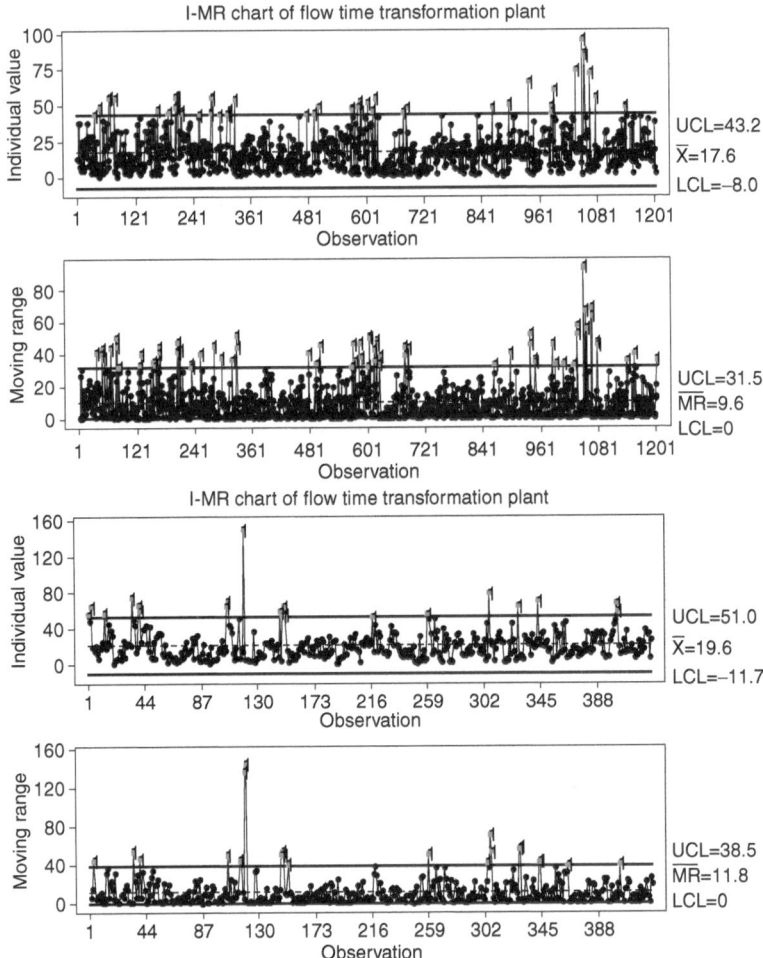

Figure 12.7   Cycle time for plant with higher process variation

As so much time was spent in WIP this led the author to believe that the WIP control, or policies around how it was made, were having the greatest effect on cycle time and the system, and that variability in cycle time was being added to the process by bad control and not by the process rate variability. To investigate, a daily measure of WIP was taken at random times. This can be seen in Figure 12.9.

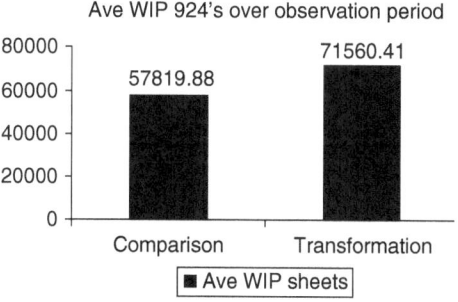

Figure 12.8 Work in Progress (WIP) in the two plants

The effect of batching at the corrugator (the supplying process to the converters) for efficiency reasons and therefore arrivals at the WIP area were causing variable flow time. The flow time was largely dependent on where in the queue a new job arrived in relation to the WIP, which could be at any point between 4 and 28 hours. The effect on flow time is seen in the previous run charts. At this point, reference can be made again to *Factory Physics* and the use of Kingman's equation for factors that determine the time spent in a queue.

Kingman's Equation

$$CT_q = \left[\frac{c^2_a + c^2_e}{2}\right]\left[\frac{u}{1-u}\right]t_e$$

The equation simply shows that the part of the cycle time spent in a queue is determined by not only the process variation of the station (the hourly rate) but also the arrival variation and the utilisation. When measuring arrivals from the corrugators to the WIP area for the converter in question, arrivals were classed as highly variable, scoring 1.7 (see Figure 12.10). Using the average arrival rate divided by the throughput rate of the converter, the machine was always 100 percent utilised (note that at 100 percent utilisation, cycle time goes to infinity in this case). When WIP lanes were full, the supplying process was shut off (although the WIP area was not always full of the right inventory!). This can be likened to the effect of Muri and Mura, i.e. uneven arrivals, high utilisation and overburden (Bicheno, 2006 p. 49). In comparison, the process variation was classed as low, scoring 0.2. In summary, we can see that by substituting these numbers into the equation, blindly using

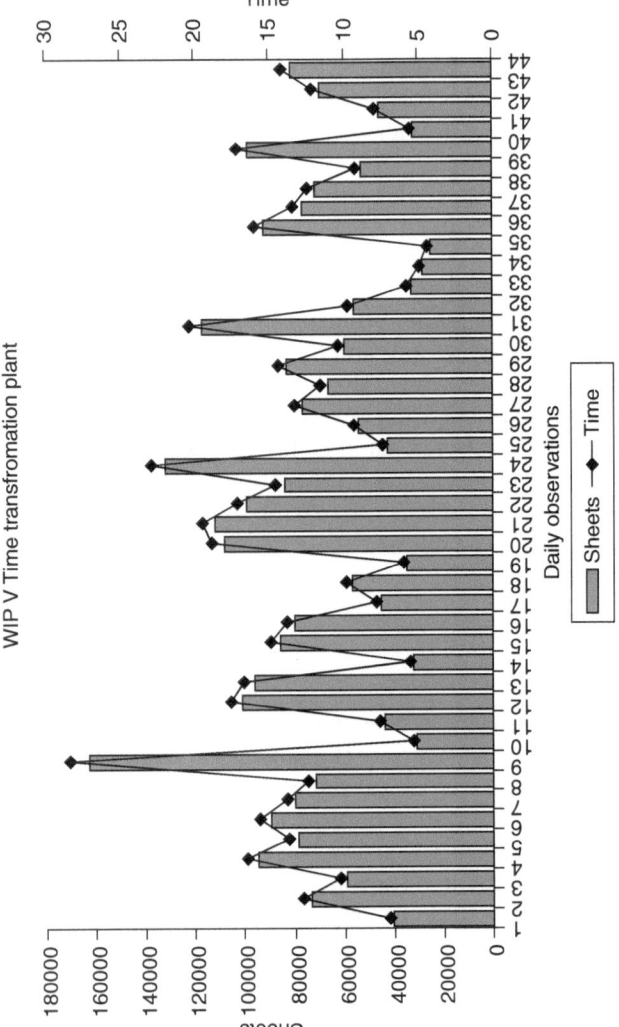

Figure 12.9 Daily measures of WIP over time

Source (Hopp and Spearman, 2000, p. 282)

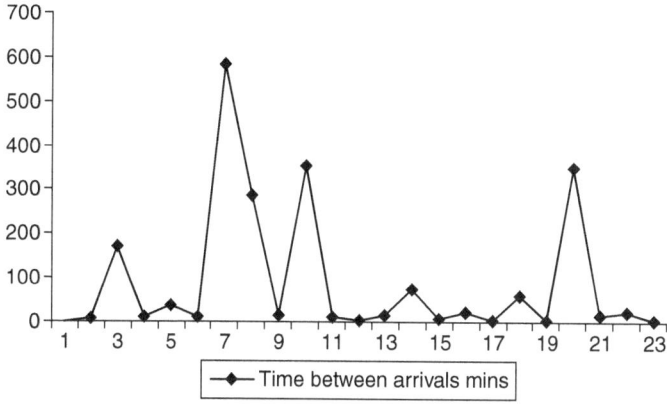

Figure 12.10 Time between arrivals at the converting process

the lean tools to reduce process variation or to increase OEE will do nothing for flow and improving the manufacturing system, as process variation in this case contributes much less towards total system time.

$$CT_q = \left[\frac{(1.7) + (0.2)}{2}\right] \left[\frac{100\%}{1-u}\right] t_e$$

In relation to Kingman's equation, with such high arrival variation and high utilisation of the machine, there is a choice to be made as to whether to work on a transformation to reduce cycle time, or to attempt to make the process more regular. The choice would no longer be to aim at improving OEE.

Even if the number for the comparison plant into the same equation for queue time is substituted (arrival variation = 1.5, process variation = 0.3, utilisation = 100 percent) this shows that the lean tools will do little to improve flow!

The study of the time between arrivals below for the transformation plant shows that the arrivals are very variable and that the flow of the product is very irregular. This is driven again by choices and policies about the most efficient way to run the plant which are mostly cost driven as they try to achieve the budgeted volume. The main thinking revolved around 'how many square metres am I making?' with no consideration of the effect on the customer. All thoughts and measures were very much focused internally. As Seddon would say, the business was guilty of 'managing the budget' and trying to benefit from 'economies of scale' rather than serving the customer and benefiting from 'economies of flow' (Seddon, 2005, pp. 22–23).

However, there was some benefit from examining the WIP graph as it can be ascertained that there are no occasions where WIP runs out and throughput is lost. This confirms another *Factory Physics* point: a pull system is only as effective as the strictness of a WIP cap. In this case, if a WIP cap was in place, then cycle time would be capped and less variable, and the system time would become more predictable. This shows that actually making more and meeting the volume targets increases the system time and decreases service level! See Figure 12.10 which illustrates this point

It was also true that once WIP arrived into the holding area, quite often it was shuffled out of FIFO (First In, First Out) sequence to save on set up at the converters (again in the hope of achieving economies of scale). This had the effect of increasing output but also increasing cycle time. At this point, there needed to be a capacity consideration: if the converters were capacity constrained against the market, then an effort should be made to maximise the throughput by reshuffling. However, this will not undo the effectiveness of having a WIP cap as cycle time will still be between a minimum and maximum. The old sequencing logic is shown in Figure 12.11.

Managing cost increases cost

The *Factory Physics* framework tells us that 'variability will be buffered in some way by inventory, capacity or time'. The variability in system time (manufacturing lead time) caused by bad control traditional measures and an internal focus on the budget, rather than process variation or

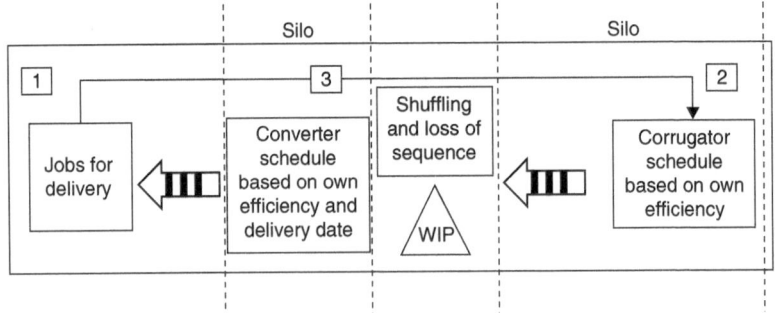

Figure 12.11 The old sequencing logic

unreliable machines was having a major effect on the plant. Customer lead times were extended as a safety buffer against the unreliable flow. The transformation plant's delivery dates, as in all MRP systems, were based on average move times and process times. It also emerged that no queuing time was included in the calculation for delivery dates and as most time was spent in a queue, delivery dates were always wrong. The logic can be seen in Figure 12.12.

This meant that over time the system for offering delivery dates was compromised. Delivery dates included a manual time buffer to cope with variability. This leads to longer than average lead times being offered, which as Goldratt suggests, decreases 'competitive edge' or increases cost.

The make-up of customer lead time following this logic is shown in Figure 12.13.

To counter this bad control and improve flow, an alternative sequencing logic was proposed. In addition to our first use of DBR principles and protecting the converter with an unlimited amount of WIP, the major change was the addition of a WIP cap between the corrugator and converter. The logic is illustrated in Figure 12.14.

Using this sequence allowed the plant to:

- Hold WIP as constant as possible (reducing arrival variation), to ensure a maximum level (WIP is bounded) and run FIFO
- Reduce manufacturing lead time, and become more predictable
- Reduce the time buffer (the safety lead time)

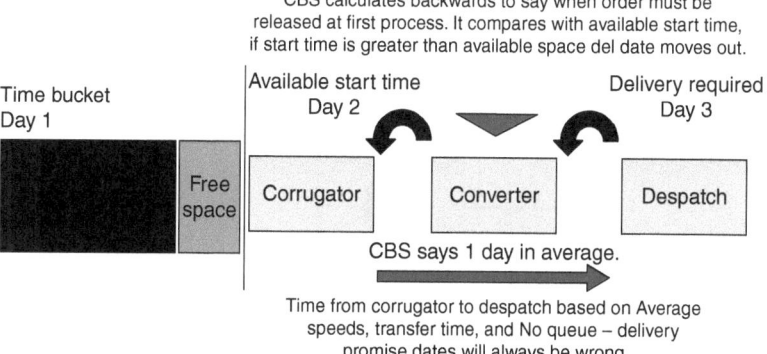

Figure 12.12 The MRP system logic for creating delivery dates

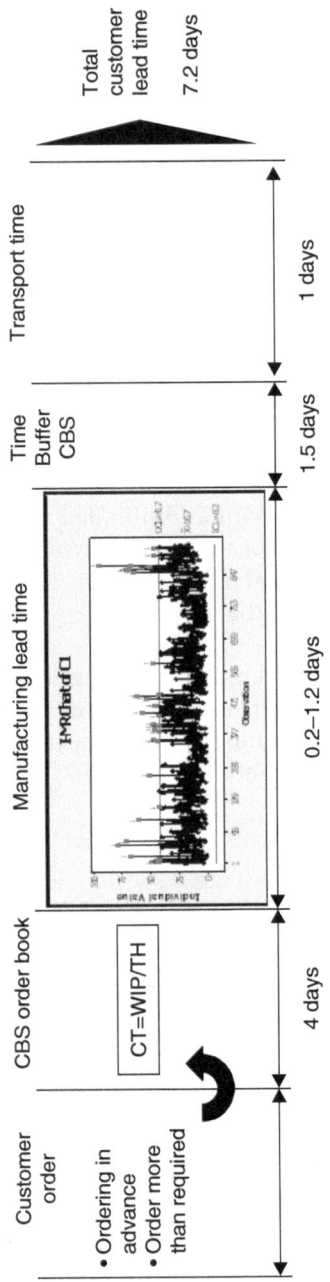

Customer order
• Ordering in advance
• Order more than required

CBS order book

$CT=WIP/TH$

4 days

Manufacturing lead time

I-MR chart of C1

0.2–1.2 days

Average 0.7 days

Time Buffer CBS

1.5 days

Transport time

1 days

Total customer lead time

7.2 days

Figure 12.13   The make-up of average customer lead time in transformation plant

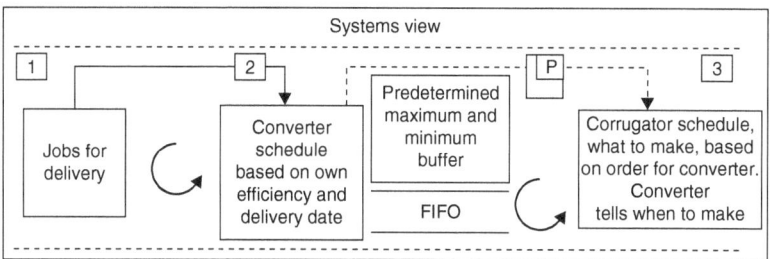

Figure 12.14 Lean transformation sequencing logic

- Reduce total lead time
- Further work to reduce the buffer and WIP without losing throughput
- Reduce manufacturing lead time even further and increase responsiveness
- Encourage customers to order correct quantities and closer to the required time
- Further reduce customer lead time
- Increase the businesses competitive edge

It was notable that no extra work was needed to improve changeover or to improve the EPE (Every Product Every scheduling interval) measure. This was simply done by choosing to run and measure differently.

Having a constant cycle time (manufacturing lead time) allows the time buffer to be reduced thus reducing total customer lead time as seen in Figure 12.15. <Figure 12.15>

These changes to the physical and policy choices were supported by measures and performance management in the scheduling department. Daily reviews are held at a visual performance board which display lead time, manufacturing lead time, average WIP, occasions over WIP maximum, and lost time for no material at the drum. The traditional measures were not completely erased but took on a lower priority.

At this point our frame of reference for the lean transformation was changing. To the question, 'How do we make this plant 'leaner' if there was little need to increase throughput?' the answer was to cut customer lead time by 50 percent and therefore increase the competitive edge of an already excellent business. This approach would now help to overcome the issue of no extra sales in the market as lead time improvement could not only be communicated but used to capture extra sales.

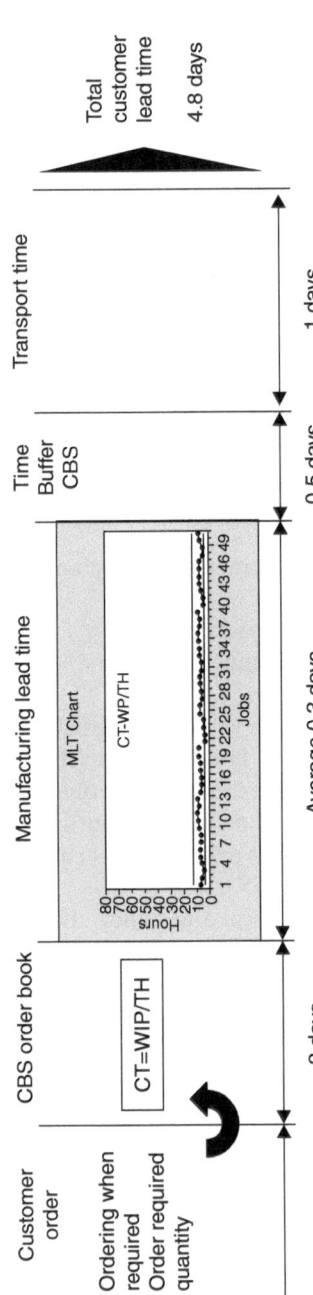

Figure 12.15 A constant cycle time allows for time buffers to be reduced

Conclusion

This work has shown the folly of applying a 'one size fits all' approach to transformation. Before starting, the team did not know exactly what the problem was that they were trying to solve, but they knew that lean tools were the solution! It was only by delving deeper and comparing the operations of two factories that it was possible to discover that the focus for improvement needed to be on creating a better flow of work.

This experience has demonstrated that variability is the key factor in designing a lean transformation path and has confirmed two clear levels for lean implementation: one which concentrates on optimising the individual functions within a factory, and one which concentrates on flow to the customer.

Before tackling improvement, the current state needs to be studied thoroughly as a system. Adding the following aspects to a current state analysis will enable a greater level of understanding: study of the measures of manufacturing flow time, the coefficient of variation for process rates and arrival rates and an analysis of variation in WIP levels (arrivals).

Where process variation is classed as medium (MV), high (HV) or not capable, then an extensive use of stability tools needs to be used as a 'first tier' approach. However, if the variation is low, as in this case, tools such as TPM (and OEE) and 5S will do little to improve flow. In the majority of the cases, process variation does not have a significant effect on flow or lost throughput. In such cases a 'second tier' approach can be used aimed at reducing variation from arrivals. Reduced cycle time and increased throughput was created by changing policies and physical factors, managing the constraint with a minimum buffer to protect throughput, and using a maximum buffer to reduce arrival variation.

References

Bicheno J. Fishbone (2006) *Flow: Integrating Lean, Six Sigma, TPM and TRIZ*. Picsie Books: Buckingham, Foreword by Kate Mackle.

Goldratt, E. M. and Cox (1984) *The Goal*. North River Press: Great Barrington, USA.

Goldratt, E. M. and Fox, R.E. (1986) *The Race*. North River Press: Great Barrington, USA.

Hopp, W. J. and Spearman, M. L. (2004) 'Commissioned Paper, To Pull or Not to Pull: What is the Question.' *Manufacturing and Service Operations Management*, 6 (2), 133–148.

Hopp, W. J. and Spearman, M. L. (2000) *Factory Physics: Foundations of Manufacturing Management*. 2nd ed. McGraw-Hill Book Company: Singapore.

Hopp, W. J. and Spearman, M. L. (1996) *Factory Physics: Foundations of Manufacturing Management*. 1st ed. IRWIN: USA.

Mackle, K. [No date]. *The Corus Way: Continuous Improvement Guide Books 6, Flow*.

Mackle, K. (2007) *MSc lecture Slides*. Cardiff University.

Ohno T. (1988) *Toyota Production System*. Productivity Press: Portland, USA.

Seddon, J. (2005) *Freedom from Command and Control: A Better Way to Make the Work Work*. 2nd ed. Vanguard Education : Buckingham, England.

Index